Joseph Priestley, John Elliot

# An Account of the Nature and Medicinal Virtues of the Principal

# Mineral Waters

of Great Britain and Ireland, and those most in repute on the continent: to which

are prefixed, Directions for impregnating water with fixed air

Joseph Priestley, John Elliot

**An Account of the Nature and Medicinal Virtues of the Principal Mineral Waters**
*of Great Britain and Ireland, and those most in repute on the continent: to which are*
*prefixed, Directions for impregnating water with fixed air*

ISBN/EAN: 9783337411046

Printed in Europe, USA, Canada, Australia, Japan

Cover: Foto ©berggeist007 / pixelio.de

More available books at **www.hansebooks.com**

A N

# ACCOUNT

OF THE

## NATURE AND MEDICINAL VIRTUES

OF THE

## Principal Mineral Waters

OF

## GREAT BRITAIN AND IRELAND,

AND THOSE

MOST IN REPUTE ON THE CONTINENT.

TO WHICH ARE PREFIXED,

Directions for IMPREGNATING WATER with FIXED AIR, in order to communicate to it the peculiar Virtues of PYRMONT WATER, and other MINERAL WATERS of a similar Nature. Extracted from Dr. PRIESTLEY's EXPERIMENTS ON AIR.

## WITH AN APPENDIX,

Containing a Description of Dr. NOOTH's Apparatus, with the IMPROVEMENTS made in it by others. And a Method of IMPREGNATING WATER with SULPHUREOUS AIR, so as to imitate the AIX-LA-CHAPELLE and other SULPHUREOUS WATERS.

## BY JOHN ELLIOT, M.D.

LONDON:

PRINTED FOR J. JOHNSON, N° 72, ST. PAUL's CHURCH-YARD, MDCCLXXXI.

# ADVERTISEMENT.

Dr. PRIESTLEY's Pamphlet on the Impregnation of Water with Fixed Air being out of print, and that Gentleman having no intention of republifhing it, I have judged proper to prefix it to the following Traꞓt, with the additions as printed in his fecond Volume of Experiments on Air. This was done as well that the reader might be entertained with the hiflory of this difcovery, as ⁻inſtruꞓed in an eafy method of making the impregnation when Dr. Nooth's apparatus might not be at hand.

J. E.

N. B. In the introduꞓion to the alphabetical lift of mineral waters, thofe of Bath, Matlock, and Chaude-Fontaine, are, after the example of preceding writers, fet down among the fulphureous waters ; but they ought rather to be claffed with the waters impregnated with fixed air. The term *calcareous Glauber's falt,* is alfo ufed in one or two places, from the fame authors, though it is not ſtriꞓly proper ; modern chemiſtry having difcovered that the bafis of that falt is not *calcareous earth,* but *magnefia.*

# E R R A T A.

Page 76, line 6 and 7, dele *Bath* and *Matlock*.

79. l. 13, dele *calcareous Glauber's*.

Errata of lefs confequence the reader will eafily correct.

OF THE

# IMPREGNATION of WATER

WITH

# FIXED AIR.

## CHAPTER I.

### *The Hiſtory of the Diſcovery.*

I T often amuſes me when I review the
hiſtory of experimental philoſophy, to
obſerve how very nearly one diſcovery is
connected with another, and yet that, for a
long time, no perſon ſhall have perceived
that connection, ſo as to have been actually
led from the one to the other; and eſpe-
cially that he who made the firſt diſcovery
ſhould ſtop ſhort in his progreſs, and not
advance a ſingle ſtep farther, to make the
other, which was perhaps of infinitely more
conſequence. And yet the caſe may be
ſuch, that it ſhall be ſo far from requiring
more genius, or ingenuity, to advance that
other ſtep, that it is rather a matter of wonder, how it was poſſible for the moſt com-

B                                           mon

mon capacity to ſtop ſhort of it. We alſo frequently find that they who make the moſt important philoſophical diſcoveries overlook the moſt obvious *uſes* of them. Several ſtriking examples of this kind will be found in my *Hiſtory of electricity,* and alſo in the *Hiſtory of diſcoveries relating to viſion, light, and colours.*

In ſuch caſes as theſe it behoves an hiſtorian to be much on his guard, leſt he ſhould haſtily conclude that to have been fact which he only *imagines* muſt have been ſo, but for which no direct evidence can be produced. As this is a caſe of ſome curioſity reſpecting the human mind, I ſhall give an inſtance of it; and I am able to produce a very remarkable one relating to the ſubject of this ſection.

When it was diſcovered that the acidulous taſte and peculiar virtues of Pyrmont water, and other mineral waters of a ſimilar nature, were owing to the fixed air which they contained; when this air had been actually expelled from the water, and it was found that the ſame water, and even other water, would

reim-

reimbibe the fame air; we are apt to conclude that the perfon who made thefe difcoveries, and efpecially the laft of them (who alfo muft have known that fixed air is a thing very eafy to be procured) muft have immediately gone to work to reduce this *theory* into *practice,* by actually impregnating common water with fixed air, in order to give it the peculiar virtues of thofe medicinal mineral waters which are fo highly, and fo juftly valued, and which are procured at fo great an expence, efpecially in this country. Accordingly, Dr. Nooth has advanced, Phil. Tranf. vol. 65, p. 59, that " the poffibility of " impregnating water with fixed air was no " fooner afcertained by experiment, than " various methods were contrived to effect " the impregnation;" and I doubt not this ingenious philofopher impofed upon himfelf in the manner defcribed above. This, however, is fo far from being the cafe, that I do not believe it is poffible to produce the leaft evidence that any perfon had the thing in view before the publication of my pamphlet upon that fubject, in the year 1772.

B 2 Indeed

Indeed had this thing been fo much as *an object of attention* to philofophers, it is impoffible but that fome of them muft have hit upon a method that would have fufficiently fucceeded. Nay, the thing is fo very eafy, and the end attainable in fo many ways, that there muft have been, in a very fhort time, a great variety of methods to impregnate water with fixed air, as there are now; and we fhould certainly have heard of *artificial mineral waters* being made according to them. It is impoffible not to conclude fo, when we confider the *time that has elapfed* fince the publication of all the difcoveries that led to it.

Dr. Brownrigg's paper, giving an account of his difcovery of fixed air in the Spa water, was read at the Royal Society June the 13th 1765, and was publifhed in 1766. This excellent philofopher compleatly decompofed that mineral water, but he gives no hint of his having fo much as attempted to *recompofe* it, or of making a fimilar water, by impregnating common water with the fame volatile principle. It is fufficiently evident that he had not thought of this, though we may wonder

wonder that he fhould not have done it, becaufe he has not mentioned it, as an object
of purfuit.

In the year following, Mr. Cavendifh's
valuable papers on the fubject of factitious
air were publifhed. He firft afcertained how
much fixed air a given quantity of water
could be made to imbibe; yet it does not
appear that he ever thought of *tafting* the
water, much lefs that he thought of making
any *practical ufe* of his difcovery.

If any negative argument can be decifive,
it is that in 1772, the very year in which my
pamphlet came out, Dr. Falconer publifhed
his excellent and elaborate treatife on the
*Bath waters*, in which he treats very largely of mineral waters in general, and all
their poffible impregnations; and yet, though
he treats of *fixed air* as one ingredient in
many of them, fee p. 185, he drops no hint
about compofing fuch water, by imparting
fixed air to common water. Alfo on the
12th of September in the fame year, Dr.
Rutherford publifhed his ingenious *Differtation on Fixed Air*, in which he fpeaks of the

prefence

prefence of it in Pyrmont water, p. 3, but
without giving the leaft hint of his being
acquainted with any method of imitating
them. And yet Dr. Nooth fays, in fact,
that from the year 1766, at the lateft, *various
methods* were contrived to effect the impregnation, though he allows that I was the
only perfon who " publifhed any defcription
" of an apparatus calculated entirely for
" this purpofe."

According to this account of the matter
there were, in the interval between 1766 and
1772, a fpace of fix years, a variety of methods for impregnating water with fixed air,
fome of them prior to, and pèrhaps much
better than mine (though he gives no hint
of his own having been invented in that
period, but fpeaks of it as fuggefted by the
confideration of the imperfection of mine)
but that I happened to get the ftart in the
publication. Dr. Falconer, however, though
the friend of Dr. Nooth (fee his treatife on
Bath Water, vol. 2. p. 323) had certainly
never heard of any of thofe methods, or
even of mine, at the very termination of
that period; and though my own acquaintance

tance with philofophical and medical people is pretty extenfive, I never heard of any
of the *various methods* that Dr. Nooth fpeaks
of; nor fince the publication of my method
have I heard of any perfon whatever having
pretended to have done the fame thing before; though nothing is more common
than fuch claims, and very often on the moft
trifling pretences.

Mr. Venelle, indeed, immediately upon
the tranflation of my pamphlet into French,
which was within a few weeks after the publication of it in Englifh (owing to the lauda-
ble zeal of Mr. Trudaine, for promoting all
philofophical and ufeful improvements) publifhed an extract of his papers from the *Me-
moires de Mathematique & de Phyfique,* to
vindicate to himfelf not my difcovery, but,
in fact, that of Dr. Brownrigg. However,
what he pretends to have difcovered was,
that the virtues of the acidulous waters were
owing to *air, in general,* without having
any idea of the difference between fixed air
and common air; fo that his difcovery was
fo far from being the fame with mine, that
it could not poffibly have led into it.

B 4 As

As I have hitherto only publifhed the method of impregnating water with fixed air
in a fmall pamphlet, for the ufe of thofe
who might chufe to reduce it into practice,
without giving any account of the manner
in which the difcovery (if it deferves to be
called one) was made, which has been my
cuftom with refpect to every thing elfe, I
fhall do it here; and I hope the narrative
will not be altogether difpleafing, as this
bufinefs has gained fo much attention in
all parts of Europe, as well as in England,
and promifes in a fhort time to fave the
very great expence of tranfporting acidulous
waters to confiderable diftances, by fuper-
feding, in a great meafure, the ufe of them.
And though what I have done in this bufinefs has certainly the leaft merit poffible
with refpect to *ingenuity,* I fhall always confider it as one of the *happieft* thoughts that ever
occurred to me; becaufe it has proved to be
of very fignal *benefit* to mankind, and will,
I doubt not, be of much more confequence
in a courfe of time.

It was a little after Midfummer in 1767,
that I removed from Warrington to Leeds;
and

and living, for the firſt year, in a houſe
that was contiguous to a large common
brewery, ſo good an opportunity produced in
me an inclination to make ſome experiments
on the fixed air that was conſtantly produced
in it. Had it not been for this circumſtance,
I ſhould, probably, never have attended to
the ſubject of air at all. Happening to have
read Dr. Brownrigg's excellent paper on the
Spa water about the ſame time, one of the
firſt things that I did in this brewery was to
place ſhallow veſſels of water within the
region of fixed air, on the ſurface of the
fermenting veſſels; and having left them all
night, I generally found, the next morning,
that the water had acquired a very ſenſible
and pleaſant impregnation; and it was with
peculiar ſatisfaction that I firſt drank of this
water, which I believe was the firſt of its
kind that had ever been taſted by man.

This proceſs, however, was very ſlow.
But after ſome time it occurred to me, that
the impregnation might be accelerated, by
pouring the water from one veſſel into another, while they were both held within the
ſphere of the fixed air; and accordingly I
found

found that I could do as much in about five
minutes in this way, as I had been able to do
in many hours before. Several of my friends
who vifited me while I lived in that houfe
will remember my taking them into that
brewery, and giving them a glafs of this
artificial Pyrmont water, made in their
prefence. Among others, I will take the
liberty to mention John Lee, Efq; of Lin-
coln's Inn, who was particularly ftruck
with the contrivance, and the effect of it.
This was in the fummer of the year 1768.

One would naturally think, that having
actually impregnated common water with
fixed air, produced in a brewery, I fhould
immediately have fet about doing the fame
thing with air fet loofe from chalk, &c. by
fome of the ftronger acids; and I do remember that it did occur to me that the thing
was poffible. But, eafy as the practice proved
to be, no method of doing it at that time
occurred to me. I ftill continued to make
my Pyrmont water in the manner abovementioned 'till I left that fituation, which was
about the end of the fummer 1768; and from
that time, being engaged in other fimilar
purfuits,

purfuits, with the refult of which the public are acquainted, I made no more of the Pyrmont water 'till the fpring of the year 1772.

In the mean time I had acquainted all my friends with what I had done, and frequently expreffed my wifhes that perfons who had the care of large *diftilleries* (where I was told that fermentation was much ftronger than in common breweries) would contrive to have veffels of water fufpended within the fixed air, which they produced, with a farther contrivance for agitating the furface of the water; as I did not doubt but that, by this means, they might, with little or no expence, make great quantities of Pyrmont water; by which they might at the fame time both ferve the Public, and benefit themfelves. For I never had the moft diftant thought of making any advantage of the fcheme myfelf.

In all this time, viz. from 1767 to 1772, I never heard of any method of impregnating water with fixed air but that abovementioned. My thinking at all of reducing to practice any method of effecting this, by air diflodged from chalk, and other calcareous fubftances,

was

was owing to a mere accident. Being at dinner with the Duke of Northumberland, in the spring of the year laſt mentioned, his Grace produced a bottle of water diſtilled by Dr. Irving for the uſe of the navy. This water was perfectly ſweet, but, like all diſtilled water, wanted the briſkneſs and ſpirit of freſh ſpring water; when it immediately occurred to me that I could eaſily mend that water for the uſe of the navy, and perhaps ſupply them with an eaſy and cheap method of preventing or curing the ſea ſcurvy, viz. by impregnating it with fixed air. For having been buſy about a year before with my experiments on air, in the courſe of which I had aſcertained the proportional quantity of ſeveral kinds of air that given quantities of water would take up, I was at no loſs for the *method* of doing it in general, viz. inverting a jar filled with water, and conveying air into it from bladders previouſly filled with air. This ſcheme I immediately mentioned to the Duke and the company, who all ſeemed to be much pleaſed with it, and expreſſed their wiſhes that I would attend to it, and endeavour to reduce it into practice; which I promiſed to do.

The

The next day I provided a fmall apparatus, adapted to this purpofe, at my lodgings, which was very eafy, as it required no other veffels but fuch as are in conftant family ufe, and with this I prefently impregnated a quantity of the New River water, fo as to make it imbibe about its bulk of air. But I was far from having hit upon the *eafieft method* of doing it; for my jars were of an equal width throughout. However, with thefe veffels the procefs was compleated in about twenty minutes, or half an hour.

A few days after this, having an invitation to wait upon Sir George Savile, I carried with me a bottle of my impregnated water, and told him the ufe that might be made of it, viz. that of fupplying a pleafant and whole-fome beverage for feamen, and fuch as might probably prevent or cure the fea-fcurvy. Sir George, with that warmth with which he efpoufes every thing that he conceives to be for the public good, infifted upon writing a card immediately to Lord Sandwich, propo-fing to introduce me to him, as having *a pro-fal for the ufe of the navy.* As I could make no objection, the card was accordingly writ-ten,

ten, and an anfwer was prefently returned from his Lordfhip, informing us that he would be glad to fee us the next day. Upon this I drew up fomething in the form of a *propofal*, which, accompanied by Sir George, I prefented to his Lordfhip, who promifed to lay it before the Board of Admiralty.

Prefently after this I had notice from the Secretary to the Board of Admiralty, that the *College of Phyficians* were appointed to examine my propofal, and to make their report of it to the Board, and an early day was fixed for me to wait upon them at their hall in Warwick-Lane; where, before a very full meeting, I produced a bottle of my impregnated water, and alfo, at their requeft, fetched my apparatus, and fhewed them the manner in which I had impregnated it. There were prefent feveral of the moft eminent phyficians in London; but both the *fcheme*, and the *objeᶜt* of it, appeared to be entirely new to every one of them; and moft of them feemed to be much pleafed with it.

Accord-

Accordingly, a favourable report was
made to the Board of Admiralty, and I was
acquainted by the Secretary, that the Cap-
tains of the two ſhips which were juſt then
failing for the South-Seas had orders to make
a trial of the impregnated water; and for
their uſe I drew out my *Directions* in writing, and ſent a drawing of the neceſſary apparatus. The method which I had now
got into was a great improvement upon that
which I had made uſe of before the College
of Phyſicians. For, in conſequence of giving more attention to it, I had, by that
time, brought it to the ſtate in which it is
deſcribed in the pamphlet.

In the mean time, I had, before I
left London, in the ſpring of that year,
made the experiment of the impregnation
of water with fixed air in the preſence of
moſt of my philoſophical acquaintance, and
their friends, both at my own lodgings,
and in other places. But upon none of theſe
occaſions did it appear that any of them had
heard of any other perſon having had the
ſame thing in view.

Laſtly,

Laftly, I will obferve, that Sir John Pringle, in his *Difcourfe on different kinds of air*
(in which he has, with the greateft exactnefs,
affigned to every perfon concerned in thefe
difcoveries their due fhare of praife) gives
no hint of his being acquainted with any
other method of impregnating water with
fixed air, than that which I had publifhed.
He certainly had not heard of any of thofe
to which Dr. Nooth alludes.

 As I have not to this day, directly or in-
directly, made the leaft advantage of this
fcheme; but, on the contrary, am juft fo
much a lofer by it as the experiments coft
me, I think it is not too much for the Public to allow me, what I believe is ftrictly my
due, *the fole merit of the difcovery;* which
with refpect to *ingenuity,* or fagacity, is
next to nothing; but with refpect to its *uti-
lity* is, unqueftionably, of unfpeakable value
to my country and to mankind.

CHAP. II.

# CHAPTER II.

DIRECTIONS *for impregnating* WATER
*with* FIXED AIR.

SECT. 1. *The Preface to the Directions as*
*first published.*

The method of impregnating water with
fixed air, of which a defcription is given in
this pamphlet, I hit upon in a courfe of
experiments; an account of which was
lately communicated to the Royal Society;
containing obfervations on feveral different
kinds of air, with only a hint of the method of combining this particular kind with
water or other fluids. Judging that water
thus impregnated with fixed air muft be
particularly ferviceable in long voyages, by
preventing or curing the fea-fcurvy, according to the theory of Dr. Macbride;
and all the Phyficians of my acquaintance
concurring with me in that opinion, I made
the firft communication of it to the Lords
of the Admiralty, who referred me to the
College of Phyficians; and thofe gentle-
men being pleafed to make a report fa-

C                    vourable

vourable to the fcheme, a trial has been ordered to be made of it on board fome of his Majefty's fhips. To make this procefs more generally known, and that more frequent trials may be made by water thus medicated, at land as well as at fea, I have been induced to make the prefent publication.

Sir John Pringle firft obferved, that putrefaction was checked by fermentation; and Dr. Macbride difcovered that this effect was produced by the fixed air which is generated in that procefs, and upon that principle recommended the ufe of *wort*, as fupplying a quantity of this fixed air, by fermentation in the ftomach, in the fame manner as it is done by frefh vegetables, for which he, therefore, thought that it would be a fubftitute; and experience has confirmed his conjecture. Dr. Black found that lime-ftone, and all calcareous fubftances, contain fixed air, that the prefence of it makes them what is called *mild*, and that the deprivation of it renders them *cauftic*; Dr. Brownrigg farther difcovered that Pyrmont, and other mineral

mineral waters, which have the fame acidulous tafte, contain a confiderable proportion of this very kind of air, and that upon
this their peculiar fpirit and virtues depend;
and I think myfelf fortunate in having
hit upon a very eafy method of communicating this air to any kind of water, or,
indeed, to almoft any fluid fubftance. In
fhort, by this method this great antifeptic
principle may be adminiftered in a variety
of agreeable vehicles.

If this difcovery (though it doth not
deferve that name) be of any ufe to my
countrymen, and to mankind at large, I
fhall have my reward. For this purpofe
I have made the communication as early
as I conveniently could, fince the lateft
improvements that I have made in the
procefs; and I cannot help expreffing my
wifhes, that all perfons, who difcover any
thing that promifes to be generally ufeful,
would adopt the fame method.

*The*

Sect. 2.  *The Directions.*

If water be only in contact with fixed
air, it will begin to imbibe it, but the
mixture is greatly accelerated by agitation,
which is continually bringing fresh particles of air and water into contact.  All
that is necessary, therefore, to make this
procefs expeditious and effectual, is first
to procure a fufficient quantity of this fixed
air, and then to contrive a method by
which the air and water may be strongly
agitated in the fame veffel, without any
danger of admitting the common air to
them ; and this is eafily done by first filling
any veffel with water, and introducing the
fixed air to it, while it ftands inverted in
another veffel of water.  That every part
of the procefs may be as intelligible as
poffible, even to thofe who have no previous knowledge of the fubject, I fhall
defcribe it very minutely, fubjoining feveral remarks and obfervations relating to
varieties in the procefs, and other things
of a mifcellaneous nature.

*The*

## *The Preparation.*

Take a glafs veffel, *a*, pl. 2. fig. 1. with
a pretty narrow neck, but fo formed, that
it will ftand upright with its mouth downwards, and having filled it with water, lay
a flip of clean paper, or thin pafteboard,
upon it. Then, if they be preffed clofe
together, the veffel may be turned upfide
down, without danger of admitting common air into it; and when it is thus inverted, it muft be placed in another veffel,
in the form of a bowl or bafon, *b*, with
a little water in it, fo much as to permit
the flip of paper or pafteboard to be with-
drawn, and the end of the pipe *c* to be
introduced.

This pipe muft be flexible, and air-tight,
for which purpofe it is, I believe, beft
made of leather, fewed with a waxed
thread, in the manner ufed by fhoe-makers.
Into both ends of this pipe a piece of a
quill fhould be thruft, to keep them open,
while one of them is introduced into the
veffel of water, and the other into the

C 3                         bladder

bladder *d*, the oppofite end of which is
tied round a cork, which muft be perforated, the hole being kept open by a
quill; and the cork muft fit a phial *e*, two
thirds of which fhould be filled with chalk
juft covered with water.

I have fince, however, found it moft
convenient to ufe a *glafs tube*, and to preferve the advantage which I had, of agitating the veffel *e*, I have *two bladders*,
communicating by a perforated cork, to
which they are both tied. For one bladder
would hardly give room enough for that
purpofe.

### *The Procefs.*

Things being thus prepared, and the
phial containing the chalk and water being
detached from the bladder, and the pipe
alfo from the veffel of water, pour a little
oil of vitriol upon the chalk and water;
and having carefully preffed all the common air out of the bladder, put the cork
into the bottle prefently after the effervefcence has begun. Alfo prefs the blad-

der

der once more after a little of the newly
generated air has got into it, in order
the more effectually to clear it of all the
remains of the common air ; and then introduce the end of the pipe into the mouth
of the veffel of water as in the drawing,
and begin to agitate the chalk and water
brifkly. This will prefently produce a confiderable quantity of fixed air, which will
diftend the bladder ; and this being preffed,
the air will force its way through the pipe,
and afcend into the veffel of water, the
water at the fame time defcending, and
coming into the bafon.

When about one half of the water is
forced out, let the operator lay his hand
upon the uppermoft part of the veffel, and
fhake it as brifkly as he can, not to throw
the water out of the bafon ; and in a few
minutes the water will abforb the air ; and
taking its place, will nearly fill the veffel
as at the firft. Then fhake the phial containing the chalk and water again, and
force more air into the veffel, 'till, upon
the whole, about an equal bulk of air has
been thrown into it. Alfo fhake the water

C 4                         · **as**

as before, 'till no more of the air can be imbibed. As foon as this is perceived to be the cafe, the water is ready for ufe; and if it be not ufed immediately, fhould be put into a bottle as foon as poffible, well corked, and cemented. It will keep, however, very well, if the bottle be only well corked, and kept with the mouth downwards.

## Obfervations.

1. The bafon may be placed inverted upon the veffel full of water, with a flip of paper between them, and then both turned upfide down together; but all this trouble will be faved by having a larger veffel of water, in which both of them may be immerfed.

2. If the veffel containing the water to be agitated be large, it may be moft convenient firft to place it inverted, in a bafon full of water, and then to draw out the common air by means of a fyphon, either making ufe of a fyringe, or drawing it out with the mouth. In this cafe, alfo,

fome

fome kind of handle fhould be faftened to the bottom of the veffel, for the more eafy agitation of it.

3. A narrow mouthed veffel is not neceffary, but it is the moft proper for the purpofe, becaufe it may be agitated with lefs danger of the common air getting into it.

4. The flexible pipe is not neceffary, though I think it is exceedingly convenient. When it is not ufed, a bent tube, *a*, fig. 2. (for which glafs is the moft proper) muft be ready to be inferted into the hole made in the cork, when the bladder containing the fixed air is feparated from the phial, in which it was generated. The extremity of this tube being put under the veffel of water, and the bladder being compreffed, the air will be conveyed into it, as before.

5. If the ufe of a bladder be objected to, though nothing can be more inoffenfive, the phial containing the chalk and water muft not be agitated at all, or with

the

the greateſt caution; unleſs a ſmall phial,
*a*, fig. 3. be interpoſed between the phial
and the veſſel of water, in the manner
repreſented in the drawing. For by this
means the chalk and water that may be
thrown up the tube *b* will lodge at the
bottom of the phial *a*, while nothing but
the air will get into the pipe *c*, and ſo
enter the water. If the tube *b* be made of
tin or copper, the ſmall phial *a* will not need
any other ſupport, the cork into which the
extremities of both the tubes are inſerted
being made to fit the phial very exactly.

6. The phial *e*, fig. 1. ſhould always be
placed, or held, conſiderably lower than
the veſſel *a*; that if any part of the mixture ſhould be thrown up into the bladder,
it may remain in the lower part of it,
from which it may be eaſily preſſed back
again. This, however, is not neceſſary,
ſince if it remain in the lower part of the
bladder, nothing but the pure air will get
into the pipe, and ſo into the water.

7. If much more than half of the veſſel
be filled with air, there will not be a body
of

of water fufficient to agitate, and the procefs will take up much more time.

8. If the chalk be too finely powdered, it
will yield the fixed air too faft.

9. After every procefs, the water to
which the chalk is put muft be changed.

10. It will be proper to fill the bladder
with water once every day, after it has been
ufed, that any of the oil of vitriol which
may have got into it, and would be in danger
of corroding it, may be thoroughly diluted.

11. The veffel, which I have generally
made ufe of, holds about three pints, and
the phial containing the chalk and water is
one of ten ounces; and I find that a little
more than a tea-fpoonful of oil of vitriol is
fufficient to produce as much air as will
impregnate that quantity of water.

12. If the veffel containing the water be
larger, the phial containing the chalk and
the oil of vitriol fhould either be larger in
proportion, or frefh water and oil of vitriol

muft

muſt be put to the chalk, to produce the requiſite quantity of air.

13. In general, the whole procefs does not take up more than about a quarter of an hour, the agitation not five minutes; and in nearly the fame time might a veffel of water, containing two or three gallons, or indeed any quantity that a perfon could well ſhake, be impregnated with fixed air, if the phial containing the chalk and oil of vitriol, be larger in the fame proportion.

14. To give the water as much air as it can receive in this way, the procefs may be repeated with the water thus impregnated. I generally chufe to do it two or three times, but very little will be gained by re-peating it oftener; fince, after fome time, as much fixed air will efcape from that part of the furface of the water which is expofed to the common air, as can be imbibed from within the veffel.

15. All calcareous fubſtances contain fix-ed air, and any acids may be ufed in order to fet it loofe from them; but chalk and oil

of

of vitriol are, both of them, the cheapeſt, and, upon the whole, the beſt for the pur-poſe.

16. It may poſſibly be imagined that part of the oil of vitriol is rendered volatile in this procefs, and ſo becomes mixed with the wa-ter ; but it does not appear, by the moſt rigid chymical examination, that the leaſt perceivable quantity of the acid gets into the water in this way ; and if ſo fmall a quantity as a fingle drop of oil of vitriol be mixed with a pint of water (and a much greater quantity would be far from making it lefs wholeſome) it might be difcovered. The experiments which were made to afcertain this fact were made with *diſtilled water,* the difagreeable taſte of which is not taken off, in any degree, by the mixture of fixed air. Otherwife, diſtil-led water, being clogged with no foreign principle, will imbibe fixed air faſter, and retain a greater quantity of it than other wa-ter. In the experiments that were made for this purpoſe, I was affifted by Mr. Hey, a furgeon in Leeds, who is well ſkilled in the methods of examining the properties of mineral waters.

17. Dr.

17. Dr. Brownrigg, who made his experiments on Pyrmont water at the fpring head,
never found that it contained fo much as one
half of an equal bulk of air; but in this method the water is eafily made to imbibe an
equal bulk.   For it muft be obferved, that
a confiderable quantity, of the moft foluble
part of the air is incorporated with the water, as it firft afcends through it, before it
occupies its place in the upper part of the
veffel.

18. The heat of boiling water will expel
all the fixed air, if a phial containing this
impregnated water be held in it; but it will
often require above half an hour to effect
it compleatly.

19. If any perfon would chufe to make
this medicated water more nearly to refemble
genuine Pyrmont water, Sir John Pringle
informs me, that from eight to ten drops
of *Tinctura Martis cum fpiritu falis* muft be
mixed with every pint of it.   It is agreed,
however, on all hands, that the peculiar
virtues of Pyrmont, or any other mineral
water which has the fame brifk or acidulous
tafte,

tafte, depend not upon its being a chalybeate, but upon the fixed air which it contains.

But water impregnated with fixed air does of itfelf diffolve iron, as the ingenious Mr. Lane has difcovered; and iron filings put to this medicated water make a ftrong and agreeable chalybeate, fimilar to fome other natural chalybeates, which hold the iron in folution by means of fixed air only, and not by means of any acid; and thefe chalybeates, I am informed, are generally themoft agreeable to the ftomach.

20. By this procefs may fixed air be given to wine, beer, and almoft any liquor whatever: and when beer is become flat or dead, it will be revived by this means; but the delicate agreable flavour, or acidulous tafte communicated by the fixed air, and which is manifeft in water, will hardly be perceived in wine, or other liquors which have much tafte of their own.

21. I would not interfere with the province of the phyfician, but I cannot entirely fatisfy myfelf without taking this opportunity

nity to fuggeft fuch hints as have occurred
to myfelf, or my friends, with refpect to
the *medicinal ufes* of water impregnated with
fixed air, and alfo of fixed air in other appli-
cations.

In general, the difeafes in which water
impregnated with fixed air will moft probably be ferviceable, are thofe of a *putrid* nature, of which kind is the *fea-fcurvy*. It
can hardly be doubted, alfo, but that this
water muft have all the medicinal virtues of
Pyrmont water, and of other mineral waters
fimilar to it, whatever they be; efpecially if
a few iron filings be put to it, to render it
a chalybeate, like genuine Pyrmont water.
It is poffible, however, that, in fome cafes
it may be defirable to have the *fixed air* of
Pyrmont water, without the *iron* which
it contains.

Having this opportunity, I fhall alfo hint
the application of fixed air in the form of
*clyfters*, which occurred to me while I was
attending to this fubject, as what promifes
to be ufeful to correct putrefaction in the
inteftinal canal, and other parts of. the
fyftem

fyſtem to which it may, by this channel, be conveyed. It has been tried once by Mr. Hey above-mentioned, and the recovery of the patient from an alarming putrid fever, when the ſtools were become black, hot, and very fetid, was ſo circumſtanced, that it is not improbable but that it might be owing, in ſome meaſure, to thoſe clyſters. The application, however, appeared to be perfectly eaſy and ſafe.

I cannot help thinking that fixed air might be applied externally to good advantage in other caſes of a putrid nature, even when the whole fyſtem was affected. There would be no difficulty in placing the body ſo, that the greateſt part of its ſurface ſhould be expoſed to this kind of air; and if a piece of putrid fleſh will become firm and ſweet in that ſituation, as Dr. Macbride found, ſome advantage, I ſhould think, might be expected from the ſame antiſeptic application, aſſiſted by the *vis vitæ*, operating internally, to counteract the ſame putrid tendency. Some Indians, I have been informed, bury their patients, labouring under putrid diſeaſes, up to the chin in freſh

D         mould,

mould, which is alſo known to take off the fœtor from fleſh meat beginning to putrify. If this practice be of any uſe, may it not be owing to the fixed air imbibed by the pores of the ſkin in that ſituation? Following the plough is an old preſcription for a conſumption, as alſo is living near lime kilns. There is often ſome good reaſon for very old and long continued practices, though it is frequently a long time before it be diſcovered, and the *rationale* of them ſatisfactorily explained.

Being no phyſician, I run no riſque by throwing out theſe random hints and conjectures. I ſhall think myſelf happy, if any of them ſhould be the means of making thoſe perſons, whom they immediately concern, attend more particularly to the ſubject. My friend Dr. Percival has for ſome time paſt been employed in making experiments on fixed air, and he is particularly attentive to the medicinal uſes of it; and from his knowledge as a philoſopher, and ſkill in his profeſſion, I have very conſiderable expectations.

CHAP.

# C H A P T E R  III.

*Of Dr.* NOOTH'*s Objections to the preceding Method of impregnating Water with fixed Air, and a Comparison of it with his own Method, both as published by himself, and as improved by Mr.* PARKER.

I can eafily forgive Dr. Nooth for his reprefenting me as having no other merit than the *firſt publication* of the method for impregnating water with fixed air, accounting for it as I have done before; but I cannot fo eafily forgive another paragraph in his paper, the tendency of which is intirely to difcredit a method, which, though it is, in fome refpects, inferior to his own, has neverthelefs its peculiar advantages : and every advantage cannot poffibly concur in any one method. He fays, p. 59, " Independent of the inconvenien-. " cies attending the procefs, there was " another objection to the apparatus, which, " with moſt people, might have confider- " able weight. The *bladder,* which formed

D 2          part

" part of it, was thought to render the
" water offenfive; and when the folvent
" power of fixed air is confidered, it will
" not appear improbable, that the water
" would be always more or lefs tainted by
" the bladder. In fome trials which I
" made with Dr. Prieftley's apparatus, it
" always happened that the water acquired
" an *urinous flavour*; and this tafte was,
" in general, fo predominant, that it could
" not be fwallowed without fome degree
" of reluctance."

That Dr. Nooth *did* produce an impregnated water which he could not fwallow
without reluctance, and even that, in the
trials to which he refers, he *generally* produced fuch water, I am far from doubting;
becaufe that might happen from various
caufes. But that the urinous flavour came
from the *bladder*, as fuch, I will venture
to fay is not poffible. For then it would
*always* have had the fame effect; and not
only myfelf have never perceived fuch a
flavour as the Doctor complains of, but
this is the only complaint of the kind
that I have hitherto heard of; though many
perfons

perfons of the moft delicate tafte, and particularly many ladies, have ufed the water
impregnated in my method for months together. Few perfons have had to do with
bladders, and fixed air confined in bladders,
more than myfelf; and yet I have never
feen any reafon to fufpect this great *folvent
power* of fixed air with refpect to them;
efpecially fo as to be apparent in the fpace
of a few minutes.

But fuppofing the fixed air to be capable of diffolving the whole bladder, and
to carry it along with itfelf into the impregnated water, no phyfician, or philofopher, will pretend to fay that it could
have any more tendency to give it an *urinous flavour*, than if it had been any other
membrane of the animal body.

Indeed, as the Doctor himfelf does not
pretend to fay that this ftrange urinous
flavour was the effect of *all* the impregnations of water made in my method, but
only in *fome* of them (though it was *generally* fo, in thofe particular trials) it is
evident, from his tacit confeffion, that it

D 3 muft

muſt have been an *accidental thing*, and could not have come from the bladder, which I ſuppoſe he made uſe of in all trials. For he has not done me the juſtice to acknowledge that, in my pamphlet, among the various methods of effecting the impregnation of water, I have deſcribed one in which no bladder is made uſe of. When the Doctor ſhall once more produce this urinous flavour (and as a new and curious experiment, it is certainly worthy of his farther inveſtigation) taking care that no carelefs ſervant ſhall have mixed any urine in the water that he calls for, I ſhall give this new objection to my procefs a farther examination. At preſent I am inclined to conſider this as an experiment of the ſervant, rather than of the Doctor himſelf.

Several perſons have thought that fixed air diſcharged from *impure chalk* gives the water that is impregnated with it a diſagreeable flavour, but this I have never obſerved myſelf; and any other calcareous matter may be uſed in my method, as well

as

as in that of Dr. Nooth, who recommends chalk, as the beft upon the whole.

I fhall conclude thefe animadverfions with doing what Dr. Nooth ought to have done before me, viz. fairly ftating the advantages and difadvantages of our two methods. His method requires *lefs fkill* in the operator, and *a lefs conftant attention*. It is alfo *more elegant* and cleanly, I mean with refpect to the *operator* ; for this does not at all affect the *impregnated water*. On thefe accounts I generally recommend and make ufe of his method myfelf, efpecially as the glaffes are made with improvements by Mr. Parker. But if Dr. Nooth be candid, he muft acknowledge that my method requires much *lefs time*, and is much *lefs expenfive* ; and therefore muft be more proper when a great quantity of impregnated water is wanted ; and efpecially when there is but little room to make it in.

My method indeed requires a conftant attendance, but I queftion whether, upon the whole, more than is neceffary to be given

to

to Dr. Nooth's method át intervals, if the
water be at all agitated; confidering that
mine does not require one-tenth part of
the time. And though my method requires fome little fkill and addrefs, it is
not fo much, but that many perfons, altogether unufed to experiments, have, to
my knowledge, fucceeded in it very well,
and have made the impregnated water in
a conftant way for their family ufe, and
without any affiftance befides what they
got from the printed directions. My apparatus cofts little or nothing, becaufe no
veffels are made for the purpofe; and both
the chalk and the acids are made to go as
far as poffible, by means of the convenient
agitation of the veffel in which they are
contained. Whereas Dr. Nooth's method
requires a peculiar and expenfive apparatus,
and more wafte is unavoidable in the ufe
of it. However, for the reafons above-
mentioned, I have never recommended my
own method for the ufe of a family fince
I have been acquainted with his.

What I have faid above is rather appli-
cable to the apparatus as it is made by
Mr.

Mr. Parker, than to that which Dr. Nooth has defcribed. For Mr. Parker's glaffes are, in my opinion, confiderably improved from thofe of Dr. Nooth. It may be faid that the improvements confift in *little things*; but little things may have great effects; and, after the difcovery of the *firft method* of accomplifhing this end, all *fubfequent methods* may be called little things; and they may be endlefsly diverfified, without any great claim of merit. I have feen feveral very ingenious methods fince the publication of mine, though none that I like fo much, upon the whole, as that of Dr. Nooth, improved by Mr. Parker.

In Dr. Nooth's apparatus, if any more air than is wanted be produced, the water will run out of the uppermoft veffel. To ufe his own words, p. 63, " Should more " air be extricated than is fufficient, in " the conduct of the procefs, to fill that " veffel, the water will run over the top " of it, and will continue to run as long " as any air afcends in the middle veffel, " or 'till the furface of the water is below " the extremity of the bent tube; and in

· " this

" this cafe the whole would be wet and
" difagreeable." But this difagreeable confequence can never happen in the ufe of
Mr. Parker's glaffes, becaufe the bent tube
in which the uppermoft veffel terminates
is made of fuch a length, that the water
expelled from the middle veffel can do no
more than nearly fill the uppermoft, and
can never run over; fo that whereas Dr.
Nooth's apparatus requires a conftant attendance, Mr. Parker's requires none. The
materials being once put into it, the procefs
will go on of itfelf, without any farther
care; unlefs the operator fhould chufe to
accelerate the impregnation by now and
then letting out the air that is not eafily
abforbed, and by agitating the water. This
I think to be a confiderable advantage gained
by a very eafy contrivance of Mr. Parker's,
overlooked by Dr. Nooth.

Mr. Parker derives another confiderable
advantage from a *channel* which he cuts in
the ftopper of his uppermoft veffel, or from
a ftopper with a hole through the middle,
which Dr. Nooth has not in his; fo that
either the operator muft be careful to take

it

it out during the effervefcence, or it will
be driven out, or fome of the veffels will
burft, to the .great danger of the by-
ftanders ; which actually happened in one
made by Mr. Parker, before he thought
of this method to prevent it. Whereas,
through the channel in Mr. Parker's apparatus, the common air eafily efcapes
from the uppermoft veffel, to make room
for the water to afcend; and when, in the
continuance of the procefs, the fixed air
rifes through the bent tube into the uppermoft veffel, it lodges upon the furface of
the water in it; and the communication
between it and the common air being fo
much obftructed, they are fufficiently feparated ; fo that even the water in the
uppermoft veffel has (if the production of
air be copious) almoft as much advantage
for receiving the impregnation, as that in
the middle veffel. This advantage Dr.
Nooth lofes.

Alfo, when he chufes to feparate the two
uppermoft veffels from the loweft, in order
to agitate the water, he muft either leave
the mouth of the uppermoft veffel open,

in

in which cafe he can hardly agitate the
water at all; or (as he prefers to do it) he
muft put the ftopper in, and confequently
admit the common air to pafs his valve,
and mix with the fixed air, which muft
greatly retard the abforption of it : whereas
Mr. Parker's veffels may be agitated with
the ftopper in, which, admitting the common air into the upper veffel, through the
channel cut in it (or through the hole of
the ftopper) permits the water to defcend
into the lower, on the furface of which
nothing but fixed air is incumbent. Should
any common air enter by the valve, which
in this cafe it hardly would, the finger of
the perfon who fhakes the veffel may eafily
be placed fo as to prevent it.

Laftly, I confider it as a valuable improvement in Mr. Parker's apparatus, that,
by means of the openings into the middle
and loweft veffels, clofed with ground
ftopples, the operator is enabled to draw
off his water, in order to tafte it occafionally, or to add to his oil of vitriol or
chalk, &c. at pleafure, without giving
himfelf

himfelf the trouble of feparating the veffels from one another for thofe purpofes.

The firft apparatus that I faw of Mr. Parker's had no *valve* at all, but only a glafs ftopple, with one or more fmall perforations, for the afcent of the air into the middle veffel. This I ftill generally make ufe of, without finding any occafion for a valve; the afcent of the fixed air fufficiently preventing the defcent of the water, as long as the procefs continues, efpecially when pounded *marble* is ufed. This fubftance Dr. Franklin recommended to me, and I give it the preference very greatly to chalk, chiefly on account of the length of time that is required to expel the air from it: For without any frefh acid, it will often continue to yield air for feveral days together.

That thofe perfons who are not poffeffed of the Englifh *Philofophical Tranfactions,* and particularly foreigners, may underftand what has preceded, I fhall give a drawing of Dr. Nooth's

Nooth's apparatus, † as improved by **Mr.** Parker, with the following general defcription of it.

In the loweft veffel, the chalk or marble, and the water acidulated with oil of vitriol, muft be put, and into the middle veffel the water to be impregnated. During the effervefcence, the fixed air rifes into the middle veffel, and refts upon the furface of the water in it, while the water that is difplaced by the air rifes through the bent tube into the uppermoft veffel, the common air going out through the channel in the ftopple. When the bent tube is of a proper length, the procefs requires no attention; and if the production of air be copious, the water will generally be fufficiently impregnated in five or fix hours. At leaft, all the attention that needs be given to it is to raife the uppermoft veffel once or twice, to let out that part of the fixed air which is not readily abforbed by water. If the operator chufe to accelerate the procefs, by agitating the water, he muft feparate the two uppermoft veffels from the loweft.

† Fig. 2.

loweft. For if he fhould agitate them all together, he will occafion too copious a production of air; and he will alfo be in danger of throwing the liquor contained in the loweft veffel into contact with the ftopple which feparates it from the middle veffel, by which means fome of the oil of vitriol might get into the water.

*End of the Extract from Dr.* PRIESTLEY's *Experiments on Air,* Vol. II.

APPENDIX.

E                        APPENDIX.

# APPENDIX.

*Dr.* Nooth's *Method of* Impregnating Water *with* Fixed Air, *as improved by Mr.* Parker, *Mr.* Magellan, *&c.*

*Defcription of the Apparatus.*

## See Fig. 4.

IT is made of glafs, and ftands on a wooden veffel *d d* refembling a tea-board, to catch any water that may chance to be fpilled, and prevent it from falling on the table. The middle veffel B has a neck which is inferted into the mouth of the veffel A, to which it is ground air-tight. This lower neck of the veffel B, has a glafs ftopple S, compofed of two parts, both having holes fufficient to let a good quantity of air pafs through them. Between thefe two parts (which may be confidered as two ftopples) is left a fmall fpace, containing a plano convex lens,

E 2                              (that

(that is, a glafs round on one fide and
flat on the other) which acts like a valve,
in letting the air pafs from below upwards,
and hindering its return into the veffel A.

The upper veffel C terminates below in
a tube *r t*, which being crooked, hinders
the immediate afcent of the bubbles of
fixed air into that veffel, before they reach
the furface of the water in the veffel B.
The veffel C is alfo ground air-tight to the
upper neck of the middle veffel B, and
has a ftopple *p* fitted to its upper mouth,
which has an hole through its middle.
The upper veffel C holds juft half as much
as the middle one B; and the end *t* of
the crooked tube, goes no lower than the
middle of the veffel B.

### The Procefs.

Fill the middle veffel B with fpring, or
any other clean and wholefome water, and
join to it again the upper veffel C. Pour
water into the veffel A (by the opening
*m*, or otherwife) fo as to cover the rifing

part

part of its bottom; About three quarters
of a pint, or a little more, will be fufficient. Fill an ounce phial with oil of vitriol, and add it to the water, fhaking the
veffel fo as to mix them well together.
As heat is generated, it will be better
to add the oil by a little at a time, otherwife a hazard is run of breaking the veffel.
Put to this, through a wide glafs, or paper
funnel, about an ounce of powdered raw
chalk, or marble *. The funnel muft be
ufed in order to prevent the powder from
touching the infide of the veffel's mouth ;
E 3                    for

* White marble being firft granulated, or pounded
like coarfe fand, is much better for the purpofe than
pounded chalk, becaufe it is harder; and therefore the
action of the diluted acid upon it is flower, and lafts a
very confiderable time. The fupply of fixed air from
it is therefore much more regular than with the chalk.
In general, it continues to furnifh fixed air more than
twenty-four hours. When no more air is produced, if
the water be decanted from the veffel A, and the white
fediment wafhed off, the remaining granulated marble
may be employed again by adding to it frefh water,
and a new quantity of oil of vitriol. A farther produce of fixed air will then be furnifhed, and this may
be repeated until all the marble be diffolved.

for if that happens, it will ſtick ſo ſtrongly to the neck of the veſſel B, as not to admit of their being ſeparated without breaking. Place immediately the two veſſels B and C (faſtened to each other) into the mouth of the veſſel A, as in the figure, and all the fixed air which is diſ-engaged from the chalk or marble by the oil of vitriol, will paſs up through the valve in S into the veſſel B. When this fixed air comes to the top of the veſſel B, it will diſlodge from thence as much water as is equal to its bulk; which water will be forced up through the crooked tube into the upper veſſel C.

Care muſt be taken not to ſhake the veſſel A when the powdered chalk is put in; otherwiſe a great and ſudden efferveſ-cence will enſue, which will perhaps expel part of the contents. In ſuch caſe it may be neceſſary to open a little the ſtopple *p,* in order to give vent, otherwiſe the veſſel A may burſt. It will be proper alſo to throw away the contents, and waſh the veſſel; for the matter will ſtick between the necks of the veſſels, and cement them together.

together. The operation muſt then be begun afreſh. But if the chalk be thrown in without ſhaking the machine; or if marble be uſed, the effervefcence will not be vio- lent: If the chalk be put into the veſſel looſely wrapt up in paper, this accident will be ſtill better guarded againſt. When the effervefcence goes on well, the veſſel C will ſoon be filled with water, and the veſſel B half filled with air; which will eaſily be known to be the caſe by the air going up in large bubbles through the crooked tube *r t*.

When this is obſerved, take off the two veſſels B and C together as they are, and ſhake them ſo that the water and air within them may be much agitated. A great part of the fixed air will be abſorbed into the water; as will appear by the end of the crooked tube being confiderably under the furface of the water in the veſſel. The ſhaking them for two or three minutes will be ſufficient for this purpoſe. Theſe veſſels muſt not be ſhook while joined to the under one A, otherwife too great an effer- vefcence will be occaſioned in the latter;

E 4                together

together with the ill confequences above-
mentioned. After the water and air have
been fufficiently agitated, loofen the upper
veffel C, fo that the remaining water may
fall down into B, and the unabforbed air
pafs out. Put thefe veffels together, and
replace them into the mouth of A, in
order that B may be again half filled with
fixed air. Shake the veffels B and C, and
let out the unabforbed air, as before. By
repeating the operation three or four times,
the water will be fufficiently impregnated.

Whenever the effervefcence nearly ceafes
in the veffel A, it may be renewed by
giving it a gentle fhake, fo that the powdered chalk or marble at the bottom may
be mixed with the oil of vitriol and water
above it; for then a greater quantity of
fixed air will be difengaged.

When the effervefcence can be no longer
renewed by fhaking the veffel A, either
more chalk muft be put in, or more oil of
vitriol; or more water, if neither of thefe
produce the defired effects.

The

The ingenious Mr. Magellan has ftill
farther improved the contrivance of Dr.
Nooth and Mr. Parker. He has two fets
of the veffels B and C. While he is fhaking the air and water contained in one of
thefe fets, the other may be receiving fixed
air from the veffel A. By this means twice
the quantity of water may be impregnated
in the fame time. He has a wooden ftand
K (Fig. 5.) to fix the veffels B C on, when
taken off from A, which is very convenient. He has a fmall tin trough for
meafuring the quantity of chalk or marble
requifite for one operation, and a wide glafs
funnel for putting it through into the veffel
A, to prevent its fticking to the fides, as
mentioned before.

He has alfo contrived a ftopple without
an hole to be ufed occafionally inftead of
the perforated one *p.* It has a kind of
bafon at the top to hold an additional weight
when neceffary. (See Fig. 6.) The ftopple
muft be of a conical figure, and very loofe;
but fo exactly and fmoothly ground as to be
air-tight merely by its preffure, which may
be encreafed by additional weights put into
                                    its

its bafon. Its ufe is to comprefs the fixed
air on the water, and thereby encreafe the
impregnation. For by keeping the air on
the water in this compreffed ftate, the latter
may be made to fparkle like Champaign.
And if the veffels are ftrong, there will be
no danger of their burfting in the operation.

If the veffels be fuffered to ftand fix or
eight hours, the water will be fufficiently
impregnated even without agitation. But
by employing the means above defcribed,
it may be done in as many minutes.

The water thus impregnated may be
drawn out at the opening *k.* But if it is
not wanted immediately, it will be better to
let it remain in the machine, where it has
no communication with the external air.
Otherwife the fixed air flies off by degrees,
and the water becomes vapid and flat; as
alfo happens to other acidulous waters.
But it may be kept a long time in bottles
well ftopped, efpecially if they are placed
with their mouths downwards.

This

This water is more pleafant to the tafte than the natural Pyrmont or Seltzer waters; as befides their fixed air, they contain faline particles of a difagreeable tafte, which are known to contribute little or nothing to their medicinal virtues; and may, in fome cafes, be hurtful. The artificial water is alfo double the ftrength of the natural; the latter containing fcarce half of the fixed air which can thus be communicated to the former.

*N. B.* Mr. Blades, of Ludgate Hill, has ftill further improved this apparatus, by changing the ftopple at *k* for a glafs cock, which is more convenient. He has like-wife altered the middle veffel B into a form more advantageous for the impregnation. See Fig. 7.

*A Method of imitating the* SULPHUREOUS *Mineral Waters, by impregnating Water with* SULPHUREOUS AIR.

We may imitate the *fulphureous* mineral waters, as well as the *acidulous* ones, or thofe impregnated with fixed air. The procefs is fufficiently fimple; and the fame apparatus will ferve for both.

Inftead of limeftone, chalk, or marble, *liver of fulphur* is to be ufed. It may be bought ready prepared of the chymifts or apothecaries; or may eafily be prepared as follows:

Mix together equal parts of brimftone, and of clean pot afhes *, and place them in a crucible, or unglazed difh, over a very gentle fire. Keep them ftirring with a ftick 'till they are united together into a blood-red mafs.

---

* Quick lime may be ufed inftead of pot afhes, taking care to chufe it well burnt.

mafs. Put it, while warm, into a bottle, which is to be kept well clofed.

Put a fufficient quantity of this fubftance, with the oil of vitriol and water, into the part A of the apparatus, and proceed as defcribed in the procefs for impregnating water with fixed air ; the *fulphureous* air will arife; the water in the middle veffel B will be impregnated with it, will fmell ftrongly ful-phureous, and refemble the celebrated waters of *Aix la Chapelle*, &c. in the fame manner as thofe impregnated with fixed air refemble thofe of *Pyrmont* and *Seltzer.*

The water thus impregnated may be heated, by putting it into a clofe veffel, placed in one that contains boiling water, and it is then a *warm fulphureous water.*

If it be not ufed immediately, it fhould be preferved in glafs or ftone bottles, well corked, and cemented, and placed with the corks downward in a cellar.

To

*To imitate more exactly the several Mineral*
*Waters.*

This confifts only in adding to the water
to be impregnated, the folid matters which
they are found to leave behind on evaporation. For example :

## I.

## PYRMONT WATER.

Add to the water in the middle veffel B, in
the proportion of about thirty grains of Epfom
falt, ten grains of common falt, a fcruple of
magnefia alba, and a dram of iron filings,
or iron wire, clean and free from ruft, to
one gallon of fpring water, and impregnate
the whole with fixed air in the manner
defcribed. Let them remain 'till the other
ingredients, and as much of the iron as is
neceffary, are diffolved, which will be in two
or three days; or the magnefia may be
omitted, and then the operation will be
finifhed in lefs than half that time.

2. SPAW

## 2.

## SPAW WATER.

Take of the foffil alkali, or fal fodæ, a fcruple, of common falt twelve grains, fpring water a gallon ; impregnate them with fixed air ; a few iron filings muft alfo be added.

## 3.

## SELTZER WATER.

Take of the foffil alkali a fcruple, common falt about the fame quantity, magnefia one fcruple, fpring water a gallon, and impregnate them with fixed air. Or it may be made without the magnefia.

## 4.

## SEIDSCUTZ PURGING WATER,
(refembling our EPSOM.)

Take of Epfom falt three ounces, water a gallon, and impregnate them with fixed air.

## 5.

## AIX-LA-CHAPELLE WATER.

Take of fea falt half a dram, foffil alkali a dram, clean chalk a fcruple, water a gallon. Impregnate them with fulphureous air.

Other waters may in like manner be imitated by adding Epfom falt for purging waters, fea falt for falt waters, &c. And as fome waters (as the cold fulphureous ones) contain both *fixed* and *fulphureous* air, a mixture of liver of fulphur and chalk may be put into the veffel A with the oil of vitriol, by which means both thefe airs will be produced, and the water of courfe impregnated with them.

AN

AN

# ACCOUNT

OF THE

## NATURE, PROPERTIES,

AND

## MEDICINAL VIRTUES

OF THE

## Principal Mineral Waters

IN

## GREAT BRITAIN AND IRELAND;

AND OF THOSE

MOST IN REPUTE IN FOREIGN PARTS.

Digefted into Alphabetical Order.

By JOHN ELLIOT, M. D.

# INTRODUCTION.

THE following treatife on Mineral Waters being intended for the Public in general, the Editor has endeavoured to couch it in fuch terms as that it may be underftood by thofe who are unacquainted with the art of phyfic. Such an account has been judged by many very proper to be fubjoined to the foregoing differtation.

All the mineral waters in *England*, of any note, will be found noticed in this tract : together with their virtues, and the method, and feafon of ufing them, fo far as could be learnt from the authors who have been confulted on the occafion. To thefe are added all the principal mineral waters of *Scotland* and *Ireland*, as well as the moft celebrated ones which the Englifh valetudinarian may have occafion to vifit on the continent.

The greateft part of the books which have hitherto been written on this fubject, abound with experiments tending to fhew the

*analyfis*

*analyſis* of thoſe waters. But this can be
of little uſe except to the faculty; and muſt
be dry, and perfectly unintereſting to common readers. Beſides, the neceſſity of ſuch
accounts is ſuperſeded by ſpecifying the ingredients themſelves with which the waters
are impregnated, and their virtues as medi-
cines; to ſhew which is the ſole end of
theſe experiments. It would alſo have
ſwelled the volume to an unwieldy ſize.
For this laſt reaſon, as alſo becauſe it was
judged wholly unneceſſary and ſuperfluous,
the deſcriptions of the places in which the
reſpective waters are ſituated, are likewiſe
omitted.

For the convenience of the reader, the
waters are arranged in *alphabetical* order, by
which means they will the more readily be
found. I wonder indeed that this method
is not obſerved by authors on many other
occaſions. For though there be a ſyſtema-
tical arrangement of the things treated of
in their books, yet the reader is, after all,
obliged to refer to an *index*; which in fact
is an alphabetical arrangement of the particulars of the ſubject.

The

The reader will find accounts of a great number of waters which he probably never heard of before. As many of thofe are of fimilar virtues to others which are more famous, the invalid will be inftructed where to find a mineral water proper for his complaint near at hand, when it might not be convenient for him, on account of the diftance, or otherwife, to repair to thofe of greater *note*, though perhaps not of fuperior *virtue*.

For this purpofe alfo, as well as for the more readily finding out waters of particular virtues, thefe waters are claffed or arranged according to their refpective mineral properties; as will prefently be feen.

Water is ufually reckoned by philofophers a fimple element; but, from the nature of the foil over which it paffes, and other accidents to which it is expofed, it is always more or lefs impregnated with foreign particles. According to the nature of thefe particles, the properties of the water are different. Hence we have *hard* water, *foft* water, *falt* water, and the almoft infinite

variety

variety of *mineral* waters. The principal
of the latter, in this part of the world,
will be found noticed in the following
tables :

### 1ſt. CHALYBEATE WATERS.

| | |
|---|---|
| Hampſtead | Peterhead |
| Carlton | Glendy |
| Iſlington | Aberbrothick |
| Leez | Cobham |
| Markſhall | Tunbridge |
| Felſtead | Buxton |
| Wellenbrow | Millar's Spaw |
| Ayleſham | Latham |
| Malvern | Tibſhelf |
| Colurian | Chippenham |
| Harrogate | Witham |
| Road | Lancaſter |
| Ilmington | Whiteacre |
| Birmingham | Weſt Aſhton |
| Cannock | Cawthorp |
| Moſs Houſe | Derby |
| Wigan | Weatherſtack |
| Sene | Filah |
| Thetford | Dortſhill |
| Lincomb | Stanger |
| Llandrindod | Dunſe |

Caſtle

| | |
|---|---|
| Caſtle Connel | Dunnard |
| Tralee | Maccroomp |
| Granſhaw | Wexford |
| Newton Stewart | Ballyſpellan |
| Galway | Nezdenice |
| Coolauran | Kuka |
| Liſdonvarna | Spaw |
| Ballycaſtle | Zahorovice |
| Glanmile | Bromley |
| Kanturk | |

Chalybeate waters are the moſt uſeful and beneficial to health of any of the mineral waters; and are very plentiful in this iſland.

Waters are known to be chalybeate by their ſtriking a reddiſh purple, or black colour with an infuſion of galls; and according to the height of the colour, provided the ſtrength of the infuſion be the ſame, we judge of the ſtrength of the water as a chalybeate.

The iron in thoſe waters is held in ſolution by means of fixed air, as may be judged from what has been already

F 4                                    ſaid

faid on this fubject. As the fixed air foon
flies off on expofing the water, the iron
falls to the bottom in form of a brownifh
yellow powder. Hence thefe waters ftrike
the deepeft black with galls at the fpring
head; and in time they wholly lofe that
property.

They have a brifk acidulous or vinous
tafte when frefh, and tinge the ftools
black.

Taken inwardly they ftrengthen the conftitution in general, encreafe the tone of
the fibres, quicken the circulation, and
reftore a proper confiftence to the blood
when in a too thin and watery ftate. And
hence they are found to invigorate the
whole frame. They are good in difeafes
arifing from weaknefs. In fpafmodic diforders, arifing from too great irritability and
relaxation of the nervous fyftem. In fluor
albus, and gleets; in female obftructions;
in hyfteric and hypochondrical diforders;
in lofs of appetite and digeftion; and in a
variety of other complaints, as will be fpeci-
fied

fied in treating of the refpective waters; they differing fomewhat in their virtues.

Previous to a courfe of thefe waters, bleeding, and a cooling purge, may be neceffary, in cafe of heat and fever; and coftivenefs fhould alfo be avoided while drinking them. Where there is much fever, and alfo in ulcers of the lungs, and in confirmed obftructions attended with fever, the ufe of thefe waters is improper.

Patients ought to begin with drinking a fmall quantity of thefe waters every morning, and gradually to increafe the dofe. A temperate and moderate diet, and gentle exercife fhould alfo be obferved while taking them.

If the waters are too cold for the ftomach, a little warm water may be mixed with them juft before drinking.

Acids, tea, and other things, which decompofe thefe waters, fhould not be taken for fome time before or after drinking them.

Befides

74       *Introduction.*

Befides iron, thefe waters ufually contain
fea falt, the foffil alkali, a purging falt, or
other fubftance, as will be noticed when
treating of them.

2d.  CHALYBEATE PURGING WATERS.

| | |
|---|---|
| Knowfley | Stoke |
| Burlington | Woodham Ferris |
| Aftrope | Hanlys |
| Coventry | Athlone |
| Bournley | Mount Pallas |
| Townley | Killinfhanvally |
| Newham Regis | Cleves |
| Binley | Driburg |
| Kingfcliff | Hoff Geifmar |
| Thirfk | Pyrmont |
| Hartlepool | Egra |
| Thornton | Nevil Holt |
| Orfton | Ballycaftle |
| Stenfield | Deddington |
| Kirby | Drig-Well |
| Tarleton | Inglewhite |
| Malton | Gainfborough |
| Afwarby | Thorp Arch |
| Scarborough | Caftlemaign |
| Cheltenham | Ballynahinch |
| Bagnigge | Jeffop |

Thefe

Thefe chalybeate waters contain a greater proportion of purging falt than of any other folid matter, and therefore when taken in fufficient quantity (feveral pints) they operate by ftool. They have this advantage over other purges, that they do not exhauft the ftrength.

If taken in lefs quantity, as alteratives, they operate chiefly by urine, and then they fall rather under the firft clafs of thefe waters than the prefent.—*See what was faid of chalybeate waters.*

## 3d. SULPHUREOUS WATERS.

| | |
|---|---|
| Sutton Bog | Mechan |
| Wiglefworth | Afhwood |
| Chadlington | Derryhence |
| Bilton | Drumafnave |
| Queen Camel | Anaduff |
| Nottington | Aphaloo |
| Drumgoon | Harrogate |
| Swadlingbar | Maudfby |
| Derrylefter | Cricklefpaw |
| Lifbeak | Broughton |
| Killafher | Shettlewood |

Reddle-

| Reddleflone | * Aix la Chapelle |
| Durham | * Borſet |
| Wardrow | * Chaude Fontaine |
| Skipton | * Bareges |
| Rippon | * Baden Baden |
| Llandrindod | * Bath |
| Moffat | * Matlock |
| Carftarphin | |

Waters called *ſulphureous* do not contain
an actual ſulphur, but are impregnated with a
gas, or ſpirit (the ſulphureous air already deſcribed) which gives them their ſulphureous
ſmell. Beſides this, they uſually contain
either the foſſil alkali, ſea ſalt, a purging
ſalt, iron, an earth, or other matter, and
commonly ſeveral of theſe in different proportions.

Waters of this ſort are diuretic, and
ſtrongly diaphoretic, and are therefore good
in cutaneous diſeaſes, uſed both internally
and externally. They are alſo good in
chronic obſtructions; and in diſorders proceeding from acidity, from worms, &c.

They uſually make ſilver appear of a
copper colour.

4th. Sul-

## 4th. Sulphureous Purging Waters\*.

| | |
|---|---|
| Aſkeron | Shapmoor |
| Croft | Upminſter |
| Cawley | Codſalwood |
| Cunley Houſe | Wirkſworth |
| Buglawton | Derrindaff |
| Loanſbury | Owen Bruen |
| Normanby | Pettigoe |

Theſe waters differ from thoſe in the preceding claſs in containing a purging ſalt as
the principal ſolid ingredient, and therefore
operating by ſtool. They are good in the
ſame diſorders as the alterative ſulphureous waters, as alſo for foulneſſes of the
bowels, &c.

## 5th. Acidulous, or Saline Waters.

| | |
|---|---|
| Seltzer | St. Bartholomew |
| Tilbury | Cape Clare |
| Clifton | Buch |
| Glaſtonbury | Tonſtein |
| Toberbony | \* Mount d'Or |
| Carrickmore | |

The

\* Some of the ch⸺⸺te purging ⸺⸺⸺ ⸺ ⸺ ⸺⸺
ſulphureous,

The waters of this clafs contain the foffil alkaline falt. This falt, as the waters are taken up from the fountain, is faturated, or rather fuperfaturated, with fixed air; hence the waters do not then manifeft any alkaline quality; on the contrary, they curdle with foap, and are termed *acidulæ.* This *fixed air,* or *aërial acid,* however, being very volatile, foon exhales when the water is heated, or ftands awhile expofed, and then the alkali manifefts itfelf.

The general virtues of thefe waters may be known from what is faid in the alphabet, under the article Seltzer Water.

6. Saline Purging Waters.

| | |
|---|---|
| Barrowdale | Dulwich |
| Leamington | Holt |
| New Cartmal, or | Stretham |
| Rougham | Kilburn |
| St. Erafmus | Moreton-fee |
| Cargyrle | Hanlys |
| Dortfhill | Conmer |
| Alford | Bagnigge |

Barnet

| | |
|---|---|
| Barnet | Seidlitz |
| North-hall | * Balaruc |
| Acton | Sea Water |
| Epfom | Dog and Duck |
| Alkerton | Kinalton |
| Ball, or Bandwell | Brentwood |
| Llandrindod | Colcheſter |
| Kenſington | Sydenham |
| Richmond | Carrickfergus |
| Upminſter | * Bagniers |

Theſe waters are impregnated with ſea ſalt, and alſo with a purging ſalt; either the *calcareous Glauber's*, or the *Epſom.*

They differ in ſtrength; ſome of them purge ſufficiently in the quantity of a pint; while two, three, four, five, or ſix pints of others are neceſſary to produce that effect. Some again are ſo weak as to require the addition of Glauber's ſalt, or other purgative.

Given in ſmall quantities they act as diuretics and alteratives.

They

They are good in fcrophulous and fcorbutic complaints, ulcers, and other difeafes
which make their appearance on the fkin,
and are likewife ufed as baths, and fomenta-
tions in thefe and other diforders.

The virtues of the preceding clafs of waters
depend in a great meafure on the prefence
of their *fixed air.* The waters of the prefent clafs feem to derive their virtues principally from the faline matters which they
contain.

## 7. VITRIOLIC WATERS.

| | |
|---|---|
| Shadwell | Hartfel |
| Weftwood | Crofs-town |
| Swanzy | Nobber |
| Haigh | Cafhmore |
| Vahls | Kilbrew. |

Thefe waters are impregnated with green
vitriol or copperas, and ftrike a black colour
with galls.

They are chiefly ufed outwardly for wafhing old fores and the like, and frequently
with

with good effect. In some cases, however, they are taken inwardly in small doses, and then they prove emetic and purgative.

8. WATERS WHICH CONTAIN AN EARTH.

| | |
|---|---|
| Newton-dale | * Briftol |
| Bale | * Buxton |
| Knarefborough | * Mallow. |
| Glavely | |

The cold waters° of this clafs have a petrifying quality. The virtues of the waters of this clafs being different, the reader is referred to the refpective articles in the alphabet for an account of them.

The above arrangement of mineral waters is intended more for the convenience of the reader not verfed in phyfic, than as a *fyftematical* one.

Had the latter idea been adopted, it would have been neceffary perhaps to have made a divifion of the waters into *hot* and *cold*, in imitation of the learned Dr. Donald Monro; from whofe ingenious work, together with thofe of Dr. Short, Dr. Rutty, and a few

others,

others, the following treatife has been
chiefly compiled. †

There are a great number of *cold* mineral
waters in England ; but the number of the
*hot* ones is very fmall.  In the above cata-
logue, the *latter* are diftinguifhed from the
*former* by having an ASTERISM placed before them.  Thofe of greateft note on the
continent however are alfo noticed ; in
many parts of which they abound.

The caufe of the heat of thofe waters is, in
fome inftances, fubterraneous fire ; as is the
cafe with fome which are fituated near vol-
canos.  In other cafes the heat arifes from the
mineral ingredients with which they are impregnated in their paffage.  And the fame may
be faid of thofe waters which are *cooler* than
the common temperature of the atmofphere.
Thus it is known, that quick-lime, the py-
rites ftone, and other fubftances, thrown into
water will make it *warm*.  On the contrary,
falts of various kinds make it *colder* than before.
The

† The quantity of waters to be drank, and fome other
particulars, are not always mentioned by authors, but
they may eafily be learnt on the fpot

The *warm* waters poſſeſs many of the virtues and properties of *cold* waters of the ſame
claſs, and which are impregnated in the ſame
manner ; but they are preferable in many
caſes, as from their warmth they are more
kindly and agreeable to the ſtomachs of weak
people, and promote perſpiration.

The warm waters are alſo uſed as warm
baths, and may in general be conſidered as
warm medicated baths; and theſe by relaxing
the fibres, are of uſe in a variety of diſorders
which take their riſe from rigidity, and from
ſpaſm, as alſo from other cauſes. Hence
their great uſe in rheumatiſms, inflammations, coſtiveneſs, &c. The cure is uſually
aſſiſted by the internal uſe of thoſe waters at
the time.

For complaints of a particular part of the
body, either the part is fomented with the
warm water, or the water is raiſed to an
height by pumps, or otherwiſe, and then let
fall with force on the diſeaſed part ; this
is called *pumping* by the Engliſh; the French
term it the *Douche.*

Contrivances are alſo uſed for raiſing theſe
waters into *vapour* or *ſteam,* and confining it

ſo

fo that it may be applied to the whole body, or to particular parts. Thefe contrivances are called *vapour baths*.

Baths are likewife made of the mud found at the bottom of thefe waters; and they have been found ferviceable in removing pains, and achs; and paralytic, and other complaints. The mud is either rubbed on the part, or the part is immerfed in it, as may be judged convenient or proper; when it is collected in quantity in a refervoir for thefe purpofes, it is called the *mud bath*.

The cold waters are alfo, in fome cafes, ufed externally.

I fhall conclude this introduction by mentioning fome of the moft obvious methods of analyzing, or difcovering the nature of mineral waters.

1. Waters are known to contain *iron*, by their exhibiting a purple or black colour with the infufion of galls; and the quantity of iron which they contain, that is, their ftrength as chalybeate waters, is determined by the deepnefs of the colour; the quantity and ftrength of the infufion being the fame. If the iron be held in folution by

fixed

fixed air, the latter will fly off on expofing the water in an open veffel, and then the iron will fall to the bottom in the form of a yellowifh or reddifh brown powder, or *oker*. But if the iron be held in folution by the vitriolic acid, in the form of green vitriol, or copperas, this precipitation of oker does not take place. The chalybeate water may alfo be known to be vitriolic by its auftere ftyptic tafte.

2. If, on the addition of fyrup of violets to a water, it turns to a *bright green*, it fhews that an alkaline falt is contained in the water, and the only alkaline falt ever found in mineral water is the *foffil*.

When this kind of water, however, is firft taken up from the well, the alkaline falt is ufually faturated with fixed air, and therefore does not change fyrup of violets green. It is frequently even *fuper-faturated* with fixed air, and therefore turns the fyrup red. But the air foon flies off on expofure, and then the effect is as above-mentioned, the water is alfo found fofter than common water, and lathers better with foap.

But

But earthy and metallic fubftances alfo change fyrup of violets green. To be certain therefore that it is the foffil alkali, add a little fixed alkaline falt, and if no powder falls to the bottom, and the water does not become turbid, it may be concluded that it is the *foffil alkali* with which the water is impregnated.

3. If on the addition of fyrup of violets a *red* colour is obferved, the water contains an *acid*. Thus, fixed air is an acid, and when waters which are ftrongly impregnated with fixed air, are firft taken up from the fountain, they are found to change fyrup of violets red. As the air flies off, however, this rednefs difappears, and if the water alfo contained an alkali, it will, on the further efcape of the fixed air, turn green, for the reafon given in the laft article.

4. If water contains *fea falt*, oil of vitriol dropt into it will caufe *white noxious fumes* to arife; which is the acid of the fea falt, whofe place the vitriolic acid had taken in the alkaline bafis of the fea falt.

If

If a fufficient quantity of the oil of vitriol be added, and the water be evaporated, not fea falt but a true Glauber's falt will remain behind.

Alfo the refiduum of thefe waters crackles and flies when placed on a red-hot iron.

5. If a mineral water contains that kind of purging falt, which the chemifts call *calcareous Glauber's*, or the *Epfom* falt, (the purging falts ufually found in mineral waters) a folution of fixed alkali added will make the water turbid, and the earthy bafis of the falt will fall to the bottom in form of a white powder.

6. Sulphureous waters are known by their fmell, and by their changing filver of a reddifh copper colour,

7. If water contains an earth deprived of fixed air, it may be difcovered by adding fixed air to it; for then the water will become turbid, and the earth will fall to the bottom. If, on the contrary, it contains an earth fuperfaturated with fixed air, * draw-

G 4                                        ing

* There is a famous water of this kind, in Rathbone-place, London.

ing part of it away by the air pump, or exposing the water to the air, or to warmth, will alfo precipitate the earth.

Waters in general contain feveral kinds of thefe folid matters, and therefore more than one of thefe methods are to be employed in detecting them.

8. Water may be known to contain fixed air by its fparkling on being poured from one veffel to another, by the explofion which follows on its being fhook in a phial half filled with it, and by the bubbles which arife when placed over the fire, long before it is hot enough to boil.

The folid matters contained in waters may alfo be known by evaporating the waters and examining their refiduums. But as this requires fome knowledge of chemiftry, it need not here be entered upon ; the defign of this treatife being rather to defcribe the medicinal virtues of mineral waters than their contents.

AN

# A N

# A C C O U N T

O F   T H E

# MEDICINAL VIRTUES, &c.

O F

# MINERAL WATERS.

## ABCOURT,

*Near St. Germains, four leagues from Paris.*

IT is a brisk chalybeate water, impregnated with fixed air, and the fossil alkali;
and resembles the waters of *Spaw* and *Ilmington*.

## ABERBROTHOCK,

*In Scotland.*

It is a chalybeate water, similar to those
of *Peterhead* and *Glendy*.

ACTON,

## A C T O N,

*Near London, in the County of Middlesex.*

The wells are much frequented in May, June, and July.

The water is clear, and without smell. But its taste is somewhat bitterish like a solution of Epsom salt.

It contains a purging salt, together with sea salt.

It is one of the strongest purging waters about London; and is noted for causing a great soreness in the fundament.

## A G H A L O O, or A P H A L O O,

*In the County of Tyrone, Ireland.*

It is a sulphureous water of the same kind with that of *Swadlingbar*, but stronger. Like that, it is also impregnated with the fossil alkali, and a small quantity of purging salt.

A I X-

# AIX-LA-CHAPELLE *,

## *In the Duchy of Juliers, Germany.*

This place has long been famous for its hot fulphureous waters and baths. They arife

* My friend, the ingenious Dr. Simmons, F.R.S. who made many experiments on the waters during his refidence at this place, has favoured me with an account of their feveral temperatures, as repeatedly observed by himfelf, with a thermometer conftrueted by Nairne.

The fpring which fupplies the Emperor's bath *(Bain de l'Empereur)* the New Bath *(Bain Neuf)* and the Queen of Hungary's bath *(Bain de la Reine de Hongrie)* ——— 127°

St. Quirin's bath *(Bain de St. Quirin)* — 112°

The Rofe bath *(Bain de la Rofe)* and the Poor's bath *(Bain des Pauvres)* both which are fupplied by the fame fpring ' ——— 112°

Charles's bath *(Bain de Charles)* and St. Corneille's bath *(Bain de St. Corneille)* ——— 112°

The fpring ufed for drinking is in the High Street, oppofite to Charles's bath; the heat of it at the pump is ——— ——— 106°

arife from feveral fources, which fupply
eight baths conftructed in different parts
of the town.

Thefe waters near the fources are clear
and pellucid; and have a ftrong fulphureous fmell refembling the wafhings of a
foul gun; but they lofe this fmell by expofure to air. Their tafte is faline, bitter,
and urinous. They do not contain iron. They
are alfo neutral near the fountain, but afterwards are manifeftly, and pretty ftrongly alkaline, infomuch that cloaths are wafhed
with them without foap.

The heat of thefe waters is upwards of
100 degrees of Fahrenheit's thermometer.

They are at firft naufeous and harfh, but
by habit become familiar and agreeable.
At firft drinking alfo they generally affect
the head.

Their general operation is by ftool and
urine, without griping or diminution of
ftrength; and they alfo promote perfpiration.

The

The quantity to be drank as an alterative, is to be varied according to the conftitution, and other circumftances of the
patient. In general, it is beft to begin
with a quarter, or half a pint in the morning, and encreafe the dofe afterwards to pints,
as may be found convenient. The water
is beft drank at the fountain. When it is
required to purge, it fhould be drank in
large and often repeated draughts.

In regard to bathing, this alfo muft be
determined by the age, fex, ftrength, &c.
of the patient, and by the feafon. The
degree of heat of the bath fhould likewife be
confidered. The tepid ones are in general
the beft, though there are fome cafes in
which the hotter ones are moft proper.
But even in thefe it is beft to begin with
the temperate baths, and encreafe the heat
gradually.

Thefe waters are efficacious in difeafes
proceeding from indigeftion, and from foulnefs of the ftomach and bowels. In rheumatifms ; in the fcurvy, fcrophula, and
difeafes of the fkin ; in hyfteric, and hypochondriacal

pochondriacal diforders ; in nervous complaints and melancholy ; in the ftone and
gravel ; in paralytic complaints ; in thofe
evils which follow an injudicious ufe of
mercury, and in many other cafes.

They ought not however to be given in
hectic cafes where there is heat and fever,
in putrid diforders, or where the blood is
diffolved, or the conftitution much broken
down.

# ALFORD, or AWFORD,

*In Somerfetfhire, about* 24 *miles fouthward
of Bath.*

This falt fpring was difcovered in 1670,
from the pigeons which flew thither in
great numbers to drink the water : thofe
birds being known to be fond of falt.

It contains a purging falt, together with
a portion of fea falt.

It is ftrongly purgative.

It

It is recommended as cooling, cleanſing, and attenuating. As a good remedy in the ſcurvy, jaundice, and other glandular obſtructions. It alſo promotes urine and ſweat, and therefore is good in gravelly and other diſorders of the kidnies and bladder; and in complaints ariſing from obſtructed perſpiration.

## A L K E R T O N,

*In Glouceſterſhire, near the City of Glouceſter.*

It is a purging water, of the nature of thoſe of *Dulwich,* and *Epſom.*

## A N A D U F F,

*In the County of Leitrim, Ireland.*

It is a ſulphureous water, of the ſame kind with thoſe of *Killaſher* and *Drumaſnave,* but weaker.

## A N T O N I A N.

See *Tonſlein.*

A S H-

## A S H W O O D,

*In the County of Fermanagh, Ireland.*

It is a fulphureous water; and contains the foffil alkali, with a fmall quantity of purging falt.

In its virtues it refembles the waters of *Drumgoon* and *Swadlingbar*.

## A S K E R O N,

*Five Miles from Doncafter, in Yorkfhire.*

It is a ftrong fulphureous water, and is flightly impregnated with a purging falt.

It is recommended internally and externally in ftrumous and other ulcers, fcabs, leprofy, and fimilar complaints.

It is good in chronic obftructions, and in cafes of worms and foulnefs of the bowels.

It operates by ftool and urine.

A S T R O P E,

## A S T R O P E,

*Near Banbury, in Oxfordſhire.*

It is a briſk, ſpirituous, pleaſant taſted chalybeate water, and is alſo gently pur-gative.

It ſhould be drank from three to five quarts in the forenoon.

It is recommended as excellent in female obſtructions, the gravel, hypochondriac, and ſimilar diſorders.

## A S W A R B Y,

*Seven Miles from Grantham, in Lincolnſhire.*

It is a fine blueiſh chalybeate water, and is gently laxative without occaſioning griping or faintneſs, or a pain in the fundament; which is a common effect of waters impregnated with ſea ſalt. In its virtues it reſembles the *Cheltenham* water.

H    ATHLONE,

## ATHLONE,

*In the County of Roscommon, Ireland.*

It is a chalybeate water, without colour or smell, but it will not keep.

It operates by urine, and is gently laxative. It seems to resemble the *Hartlepool* water.

## AYLESHAM,

*In the County of Norfolk.*

It is a light chalybeate water, similar to that of *Islington*.

## BADEN,

*In Austria, Germany.*

The waters are warm and sulphureous, and have been recommended in those disorders in which the *Bareges* and *Aix-la-Chapelle* waters have been found serviceable. They are particularly spoken of for the cure of gun-shot wounds, and the complaints which remain after them.

BADEN

# BADEN BADEN,

*In Swabia, Germany.*

There are a number of hot fulphureous fprings and baths in and near this place, which have the fame general virtues as thofe of *Aix-la-Chapelle* and *Baréges.* Taken inwardly they are alfo gently laxative.

# BAGNIGGE WELLS

## PURGING WATER.

It is fituated on the north-eaft fide of London, near Iflington, and is much frequented in the fpring.

It is a falt purging water.

Its virtues are fimilar to thofe of Pancras and Acton.

The dofe is from a pint to a quart. But it is ufually quickened with Glauber's, or other falts.

The

## The CHALYBEATE-WATER.

It is clear when it comes from the pump, and has a flight irony tafte.

When firft taken to the quantity of three or four glaffes, it is ufually purgative. But this effect does not continue after the inteftines are cleared of their vitiated contents.

In its virtues it refembles the *Orfton* and other fimilar chalybeates.

## B A L E M O R E.

See *Ilmington*.

## B A L L, or B A N D - W E L L,

### *In Lincolnfhire.*

It refembles the *Dropping-Well* water. Four or five half pints are reckoned a fufficient dofe.

## B A L A R U C,

### *In Languedoc, France.*

The waters of this place are hot, and gently purgative. They have been ufed in
many

many diforders for which falt purging waters are prefcribed.

As they are hot, they have alfo been found particularly ufeful in cafes where warm baths are proper, to affift the operation of fuch waters.

Hence they have been found particularly ufeful in palfies and rheumatifms, in fcrophulous, and many other diforders.

## BALLYCASTLE,

*In the County of Antrim, Ireland.*

It is a chalybeate water, fomewhat of the nature of thofe of *Iflington* and *Hampflead*; only it has a fœtid fmell.

## BALLYNAHINCH,

*In the County of Down, Ireland.*

It is a very clear, cold, chalybeate and fulphureous water, and is good in fcorbutic and cutaneous difeafes, in lofs of appetite, &c.

BALLY-

# BALLYSPELLAN,

*Eight Miles from Kilkenny, in Ireland.*

It is a light chalybeate water, fimilar to thofe of *Iflington*, and *Hampflead*.

# BAGNIERS,

*In the Bigorre, France.*

At this place are a variety of warm baths, which are ufed in the fame diforders as thofe of *Aix-la-Chapelle*.

The waters of fome fprings taken internally are diuretic, and others purgative.

# BAREGES.

*In the Bigorre, France.*

There are feveral fprings of hot fulphureous water at this place, which form four baths. *

The water is at firft clear ; but by ftanding throws up a thin pellicle, refembling a fine

---

* Dr. Simmons informs me, that on plunging his thermometer into the hotteft fpring the mercury rofe to 112°.

fine light oil. It has a flight fulphureous fmell, like that of eggs boiled hard. It has a foft and fomewhat naufeous tafte, and feels foft, like foap-water, or oil. Its volatile parts fly off on expofure to the air; and it is beft drank at the fountain head.

This water operates by perfpiration, and by urine; but feldom by ftool. The dofe is ufually a quart, or three pints.

It is alfo ufed as a bath; as a fomentation; and as a douche.

The Barages waters have been recommended in a variety of diforders; in rheumatifms, palfies, convulfions, cutaneous eruptions, the gout, fcurvy, &c. Alfo in wounds, ulcers, hard tumours; and they are faid to have been efficacious in old gunfhot wounds, and in hard knots in the urethra after venereal complaints.

## BARNET and NORTH-HALL.

The *former* fpring is fituated at Eaft Barnet in Hertfordfhire.

H 4 The

The *latter* lies about three miles north of High Barnet.

They are both purging waters, somewhat of the nature of *Epſom* water, but much weaker. That of *Barnet* is the ſtrongeſt of the two.

## BARROWDALE.
*The Spring is about three Miles from Keſwick in Cumberland.*

It is a ſalt water, and much of the nature of that of the ſea, but ſtronger.

It is a briſk and rough purge even to ſtrong conſtitutions, occaſioning great thirſt, and heating the body. A pint is uſually ſufficient for a doſe.

Taken in leſs quantity (half, or a quarter of a pint) it operates by urine.

It is of excellent uſe in ſcorbutic complaints, in the King's evil, and the leproſy. It is alſo powerful in removing chronic obſtructions; in clearing the blood

of

of acrimonious humours; in difeafes in the
fkin; and in all thofe complaints in which
fea water is ferviceable. Like that alfo it
may be ufed externally by way of fomentation or bath. See *Sea Water.*

# B A T H,

*In Somerfetfhire.*

This place has long been famous for its
warm chalybeate waters. There are feveral
fprings, but their waters are all of the fame
nature. There are fix baths, but the principal are the *King*'s bath, the *Queen*'s bath,
and the *Crofs* bath. The others are only
appendages to thefe. They are not equally
hot.

The water when viewed in the baths has
a greenifh, or fea colour: but in a vial it
appears quite tranfparent and colourlefs, and
it fparkles in the glafs.

It has a very flight faline, bitterifh, and
chalybeate tafte, which is not difagreeable,
and fometimes fomewhat of a fulphureous
fmell; but this latter is not ufually perceivable,

ceivable, except when the baths are filling.

As it rifes from the pump, it contains
fixed air, or other volatile acid, in a fufficient quantity to curdle milk and diffolve
iron. It is alfo flightly impregnated with
fea falt.

The Bath water operates powerfully by
urine, and promotes perfpiration. If drank
quickly, in large draughts, it fometimes
purges; but if taken flowly and in fmall
quantity, it rather has the contrary effect.
An heavinefs of the head, and inclination
to fleep, are often felt on firft drinking it.

This water when taken inwardly gives
a brifk ftimulus to the nerves and fibres,
and feems to give new life and vigour to
the whole frame. It alfo powerfully corrects putrefcent acrimony. Hence when
taken into the ftomach it is faid to dilute
and blunt whatever putrefcent humours it
meets with; while its brifk, volatile chalybeate principles ftimulate and increafe the
tone of the ftomach and bowels, and brace

up

up their fibres and nerves. Entering the circulation, they pervade the minuteſt veſſels; dilute, blunt, and correct thoſe fluids in the blood which are too putreſcent; encreaſe the action of the whole vaſcular ſyſtem to promote the circulation through the ſmalleſt veſſels; to break down groſs humours; to remove obſtructions; and to promote ſecretions of the ſkin and kidnies; for carrying off thoſe fluids that are unfit to circulate longer in the general maſs. And hence it is that they have been found ſo ſerviceable in ſuch a variety of diſorders. In female complaints, for example; ſuch as obſtructions of the menſes; barrenneſs proceeding from obſtruction and relaxation of the womb; the fluor albus, &c. In hyſteric and hypochondriacal diſorders; in complaints of the ſtomach and bowels proceeding from weakneſs and laxity, or from putreſcent humours. In pains of the ſtomach, attended with bad digeſtion, and in many cholicky and other diſorders of the ſtomach and bowels. In diſorders of the head and nerves; ſuch as palſies, epilepſies, convulſions, &c. In diſeaſes of the ſkin; the ſea ſcurvy; leproſy. In obſtructions of the

liver,

liver, fpleen, and other bowels; in gouty and rheumatic complaints; in the ftone and gravel; and in many other difeafes.

Thefe waters being of an heating na-ture, it is ufual, previous to a courfe of them, to cool the body by gentle purges, by a low diet, and, if found neceffary, by bleeding.

They may be drank from half a pint, to two, three, or four pints in a day, accord-ing to circumftances. The beft method is to take one, two, three, or four half glaffes at proper intervals in the morning; a glafs or two an hour before dinner; and as much about the fame time before fupper. The patient in the mean time fhould live upon a light diet, eafy of digeftion; to ufe proper exercife; to go early to bed; and rife be-times in the morning.

In fome cafes, however, thefe waters are hurtful. In hectic fevers, for example; in fuppurations of the lungs; in fits of the gout; and in the rheumatifm if inflamma-tory; and indeed in all cafes of inflammation; as alfo where the action of the fibres is already

ready too ftrong, the animal heat too great, and the blood thick and fizy.

The quantity of the waters drank in a day fhould be gradually encreafed to as much as the patient can bear; and after continuing that quantity a fufficient time, it fhould be as regularly diminifhed. The courfe may be continued for a month or fix weeks.

The ufual feafon for the Bath waters is in April, May, and June; and in Auguft, September and October.

Thefe waters are alfo ufed externally in a variety of diforders, and with good effect; and that either by bathing or pumping, as occafion may require, efpecially if ufed inwardly at the fame time. Forefts of crutches left there, are an ample teftimony of the efficacy of bathing in paralytic cafes. By foftening and relaxing the parts, and at the fame time giving a gentle ftimulus, they are alfo of fervice in removing many inveterate gouty and rheumatic complaints. In difeafes of the limbs, &c. arifing from obftructions; in fprained, relaxed, and ftiff joints; in

fcorbutic

fcorbutic and cutaneous difeafes, old fores
and ulcers, and in many other cafes; and
when the complaint is local, pumping is
generally preferred to bathing.

It is a certain effect of thefe and
other baths, to throw out a rednefs and
kind of eruption on the fkin, efpecially
in thofe who are fcorbutic, &c. But
this effect difappears by their continued ufe,
and the diforders themfelves are at length
cured.

The mud and fcum of thefe waters have
alfo been applied with good effect by way
of poultice in hard fwellings, in weak joints,
in contractions of the limbs, in fcald heads,
running ulcers, &c. and herbs are fometimes boiled with them in the Bath water
to a proper confiftence, for thefe and the
like purpofes.

### BILTON,

*Near Knarefborough, in Yorkfhire.*

The water has a ftrong fulphureous fmell,
and taftes fomewhat faltifh. It is colder
than common water.

It

It contains the native alkaline falt, with a little fea falt.

It acts as a gentle purge; and is fomewhat fimilar in virtue to the *Sutton Bog* water.

## B I N L E Y,

*Near Coventry, in Warwickſhire.*

It is a chalybeate water, and alfo purgative and diuretic. It refembles the *Scarborough* water, but is lefs purgative.

## B I R M I N G H A M,

*In Warwickſhire.*

Near to this place is a briſk chalybeate water, which feems to refemble that of *Hampſtead* in *Middlefex.*

## B O R S E T. ‡

*About a Mile and half from Aix-la-Chapelle, Germany.*

The waters are warm, and of the nature of

‡ The waters at this place, which is only about a mile from Aix-la-Chapelle, are diftinguifhed into the upper

and

of thofe of Aix-la-Chapelle; but they are only ufed as baths, for the difeafes in which the waters laft mentioned are recommended, and alfo in dropfical and oedematous cafes.

# B R A B A C H,

*In the Diftrict of Mengerfkirchen, in the Country of Naffau, Germany.*

It is a brifk fpirity chalybeate water, which may be preferved long in well-ftopt bottles, though it foon fpoils in the open air. It has a fomewhat falt, fulphureous, and aftringent tafte, and contains the foffil alkali.

It refembles the German *Spa Water* in its general virtues.

BRAN--

and lower fprings. The former were found by Dr. Simmons to raife the thermometer to 158°; the latter to only 127°. All the baths are fupplied by the firft. Dr. Simmons obferved that thefe waters were much lefs fulphureous than thofe of Aix-la-Chapelle, probably on account of their greater heat. He likewife found that they abounded much with felenitis, which incruft the pipe through which the water paffes, and likewife the fides of the bath.

# B R A N D O L A,

## *In Italy.*

It is a flight chalybeate water, extremely limpid and chryftalline, impregnated with an alkaline falt, and abounding in fixed air. It fmells fomewhat fulphureous, and has an acidulous tafte.

It is commonly drank from two pints to a gallon or more in a day. It promotes urine and perfpiration, and is gently laxative.

Its virtues feem to refemble thofe of the *Iflington* and *German Spaw* waters.

# B R E N T W O O D,

## *In Effex.*

It is a purging water, of the nature of thofe of Pancras, Epfom, and Dulwich.

# B R I S T O L,

## *In Somerfetfhire.*

The fprings are known by the name of the *Hot Wells.*

I                         The

The water at its origin is warm, clear, pellucid and fparkling; and if let ftand in a glafs, covers its infide with fmall air-bubbles. It has no fmell, and is foft and agreeable to the tafte. It raifes the thermometer from about feventy to eighty degrees. It contains an earthy matter, which is fufpended by means of fixed air, together with fea falt, and a fpecies of Glauber's falt. The quantities of thefe latter ingredients however are very fmall.

It has been recommended in a variety of diforders. In confumptions and weaknefs of the lungs; in cafes attended with hectic fever and heat (in which, among other properties, they differ from the *Bath* water) in uterine and other internal hæmorrhages, and in immoderate difcharge of the menfes; in old diarrhæas and dyfenteries; in the fluor albus; in gleets; in the diabetes; and in other cafes where the fecretions are too much increafed, and the humours too thin; in the ftone and gravel; in the ftrangury; in colliquative fweats; in fcorbutic and fimilar cafes; in cholics; in the gout

and

and rheumatifm ; in lofs of appetite and indigeftion ; and in many other difeafes.

The ufual method of drinking the water is a glafs or two before breakfaft, and about five in the afternoon. The next day three glaffes before breakfaft, and as many in the afternoon ; and this is to be continued during the patient's ftay at the Wells. A quarter or half an hour is allowed between each glafs.

A courfe of thefe waters requires no pre-paration further than to empty the bowels by fome gentle purge ; and if heat or fever requires, to take away a few ounces of blood. Coftivenefs, however, fhould be avoided during the courfe.

Externally they are ufeful in fore and in-flamed eyes ; in fcrophulous and cancerous ulcers ; and in other fimilar cafes.

This water is cooling and quenches the thirft. It is beft drank at the fpring head ; though it will bear carriage tolerably well.

BROM-

## B R O M L E Y,

*In Kent.*

It is a chalybeate water, refembling thofe of *Spaw*, *Iflington*, and *Hampftead*.

## B R O U G H T O N,

*In the Weft Riding of Yorkfhire, near Coln, in Lancafhire.*

It is a ftrong fulphureous water; it turns filver and copper black; it reddens the leaves of trees, &c. and makes the bottom of its bafon black.

It is impregnated with fea falt, and a purging falt; and its virtues are fimilar to thofe of the *Harrowgate* water.

## B   U   C   H,

*Situated about a German Mile from the Caroline baths in Bohemia.*

The waters have a brifk pungent tafte, and are plentifully impregnated with *fixed air*. This, on expofure, flies off, and they become infipid. In this they differ from

*Seltzer*

*Seltzer* water, which acquires a lixivial tafte by ftanding.

They contain, however, an alkaline falt; and therefore their virtues are fimilar to thofe of the *Tilbury* and *Seltzer* waters, but much weaker.

## BUGLAWTON,
### *Near Congleton, in Chefhire.*

It is a fulphureous water, impregnated with a purging falt, and in its virtues feems to refemble the *Afkeron* water.

It is intenfely cold, and has a pretty ftrong fulphureous fmell and tafte.

## BURLINGTON,
### *In Yorkfhire.*

It is a brifk chalybeate water, and re-fembles thofe of *Scarborough* and *Cheltenham,* tho' it feems to be lefs purgative.

## BOURNLEY, or BURNLEY,
### *In Lancafhire.*

It is a chalybeate water of the nature of the *Scarborough,* but lefs purgative.

# B U X T O N,

*In Derbyſhire.*

This is a hot water, reſembling that of *Briſtol.*

It has a ſweet and pleaſant taſte.

It contains the calcareous earth, together with a ſmall. quantity of ſea ſalt, and an inconſiderable portion of a purging ſalt. But no iron can be diſcovered in it.

This water taken inwardly is eſteemed good in the diabetes; in bloody urine; in the bilious cholic; in loſs of appetite, and coldneſs of the ſtomach; in inward bleedings; in atrophy; in contraction of the veſſels and limbs, eſpecially from age; in cramps and convulſions; in the dry aſthma without a fever; and alſo in barrenneſs.

Inwardly and outwardly it is ſaid to be good in rheumatic and ſcorbutic complaints; in the gout; in inflammation of the liver and kidnies, and in conſumptions of the lungs; alſo in old ſtrains; in hard callous tumours; in withered and contracted limbs;

in

in the itch, fcabs, nodes, chalky fwellings, ring worms, and other fimilar complaints.

Befides the hot water, there is alfo a cold *chalybeate* water, with a rough irony tafte. It refembles the *Cawthorp* water.

## C A N N O C K.

### *Near Stafford.*

It is one of the beft and lighteft chaly-beate waters in Staffordfhire. In its virtues it refembles thofe of *Hampftead* and *Ifling-ton*.

## C A P E  C L E A R.

### *Situated in the moft fouthern part of Ireland.*

It is a fmooth, faltifh water, and lathers with foap.

It contains an alkaline falt, together with a little fea falt.

Its virtues are fimilar to thofe of the waters of *Tilbury* and *Clifton,* but weaker.

I 4  CARGYRLE.

## C A R G Y R L E,

### *In Wales.*

The fpring is fituated about ten or twelve miles from Chefter.

The water is of the nature of the *Bar-rowdale* water, but much weaker, feveral quarts being required to be taken for a purge.

## C A R O L I N E   B A T H S,

### *At Carlfbad, in Bohemia, Germany.*

The waters of this place are hot, and are impregnated with the foffil alkali.

They are recommended externally and internally in female obftructions. In relaxed habits. In glandulous obftructions. In diforders arifing from vifcid fluids, and in a variety of other complaints; and it is faid, that they may be drank, and bathed in, by perfons of all ages and conftitutions, with fafety.

CARLTON,

# CARLTON,

*Near Newark upon Trent, in the county of Nottingham.*

It is chalybeate water, refembling thofe of *Iflington* and *Hampftead*, but it has a fœtid fmell, like infufion of horfe-dung.

# CARRICKFERGUS,

*In the county of Antrim, Ireland.*

The water is of a bluifh colour, and a very foft tafte at the fountain head.

It is weakly purgative; and muft be drank to the quantity of two or three quarts.

# CARRICKMORE,

*In Ireland.*

It is fituated about five miles from Bel-turbet, in the county of Cavan.

The water has a foft, milky tafte, like Briftol water; and putrifies by keeping.

It

It curdles a folution of foap; and with falt of tartar gives a white fediment.

It contains an alkaline falt, together with a purging falt.

Its virtues therefore are fimilar to thofe of *Tilbury* and *Clifton.*

## CARSTARPHIN.
*Two miles from Edinburgh, Scotland.*

It is a weak fulphureous water, flightly impregnated with fea falt.

There is another fpring, about a mile from Edinburgh, on the banks of the water of Leith.

They refemble the *Moffat* water in virtues; and the latter is reckoned the ftrongeft.

## CASHMORE.
*In the County of Waterford, Ireland.*

It is near the *Crofs-town* water, which it refembles in virtues, though ftronger.

CASTLE-

## CASTLECONNEL,

*In the County of Limerick, Ireland.*

It is a chalybeate water of confiderable repute, and refembles the German *Spaw* waters.

## CASTLEMAIGN,

*In the county of Kerry, Ireland.*

It is a fulphureous, and ftrongly chalybeate water, and in its virtues feems to refemble that of *Deddington.*

## CAWLEY,

*Near Dranefield, in Derbyfhire,*

It is fulphureous, and gently purgative; and refembles the *Afkeron* water.

## CAWTHORP,

*Four miles from Bourne, in Lincolnfhire.*

It is a faltifh chalybeate water, and foams much as it rifes from the fpring.

It refembles the *Tunbridge* water in virtues, but is faid to be more purgative; and is alfo a good corrector of acidities.

CHAD-

## CHADLINGTON,

*Near Chipping-Norton, Oxfordſhire.*

The water has a ſaltiſh taſte, and ſmells like the waſhings of a foul gun.

It is one of the waters termed ſulphureous.

It contains alſo the foſſil alkaline ſalt, together with a little ſea ſalt.

It acts as a purgative; and its virtues reſemble thoſe of the *Sutton Bog* water.

## CHAUDE FONTAINE.

*About two leagues from Liege, and three from Spaw, in Germany.*

The water of theſe ſprings is hot, and ſupplies fifty baths.

It is claſſed by authors with the ſulphureous waters; but Dr. Simmons, ‡ who ſpent ſome time at this place in 1776 and 1777, informs me they have no ſulphureous ſmell;

‡ The ſame gentleman informs me, that on the 5th of July 1777, when the mercury in his thermometer, in the ſhade, ſtood at 75°, it roſe in the bath to 92°.

finell; that they are impregnated with calcareous earth, and foffil alkali, and alfo with
fixed air.

They are not chalybeate; and therefore
rather refemble our *Briftol* and *Buxton* than
the *Bath* water.

Their virtues *externally* however may be
collected from what has been faid of the
*Aix-la-Chapelle* and *Bath* waters.

## CHELTENHAM.

*In Gloucefterfhire, fix miles from Gloucefter.*

It is one of the beft and moft noted
purging chalybeate waters in England, tho' it
is not fo much frequented as formerly.

The dofe is from one pint to three or
four. It operates with great eafe, and is
never attended with gripings, tenefmus, or
ftraining at ftool. It is beft taken a little
warm.

It alfo creates an appetite; is excellent in
fcorbutic complaints, and has been ufed
with fuccefs in the gravel.

CHIPPEN-

# C H I P P E N H A M,

## *In Wiltſhire.*

It is a pretty ſtrong chalybeate water, reſembling thoſe of *Iſlington* and *Tunbridge.*

# C L E V E S,

## *In the Duchy of Cleves, Germany.*

It is a briſk chalybeate water, and operates by urine. It reſembles the *Pyrmont* water.

# C L I F T O N,

## *This is a Village near Deddington in Oxfordſhire.*

The well is about a furlong ſouth of Clifton. The water is clear, and has but little taſte.

The principal ingredient contained in it is the foſſil alkali.

Its virtues are ſimilar to thoſe of the *Tilbury* water, though in a leſs degree. But as it alſo contains a purging ſalt, it is more purgative than that.

It

It has been much ufed by way of bath in diforders of the fkin.

## C  O  B  H  A  M,

*In the County of Surry.*

It is a chalybeate water, of the nature of that of *Tunbridge,* but rather ftronger of the iron.

## C O D S A L W O O D,

*Five Miles from Wolverhampton, Stafford-fhire.*

It is a ftrong fulphureous water.

In its virtues it feems to refemble the *Afkeron* water.

## C O L C H E S T E R,

*In the County of Effex.*

It is a purging water of the nature of thofe of *Acton* and *Epfom.*

## C O L U R I A N,

*In the Parifh of Ludgvan, in Cornwall.*

It is a chalybeate water, and feems to refemble thofe of *Hampftead* and *Iflington.*

COM-

## C O M N E R,  or  C U M N E R,

*In Berkſhire, four miles weſt of Oxford.*

The water is of a whitiſh colour, eſpe-cially in the ſummer.

It is purgative, and may be drank to the quantity of one, two, or three quarts, ac-cording to the patient's conſtitution.

## C O O L A U R A N,

*In the county of Fermanagh, Ireland.*

It is a chalybeate water, reſembling that of *Peterhead*, but weaker.

## C O V E N T R Y,

*In Warwickſhire.*

It is a chalybeate and purging water, which ſits eaſy upon the ſtomach, ſoon paſſes off, raiſes the ſpirits, and creates an appetite.

In its general virtues it reſembles the *Scarborough* and *Cheltenham* waters.

## C R I C K L E   S P A W,

*Situated near Broughton, in Lancaſhire.*

It is a ſtrong ſulphureous water, impreg-nated with ſea ſalt.

It

It is purgative; and in its virtues resembles the *Harrogate* water.

## C R O F T,

*In the North Riding of Yorkſhire, on the
confines of Durham.*

This is a ſtrong ſulphureous water, weakly
impregnated with a purging ſalt.

It is clear and ſparkling, and its ſtream
does not riſe or fall by rain or drought.

It is purgative, and of the nature of the
*Aſkeron* water; and is ſaid to have per-
formed remarkable cures.

## C R O S S - T O W N,

*Near the Town of Waterford, Ireland.*

It reſembles the *Hartfell* water in Scotland.

This water vomits ſome, purges others,
and with others operates by urine.

It ſeems at ſome times to contain a greater
quantity of acid than at others.

K             CUNLEY

## CUNLEY HOUSE,

*Near Whaley, in Lancashire.*

It is ftrongly fulphureous, and gently purgative, and feems to refemble in its virtues the *Afkeron* water.

## DAS WILD-BAD,

*Within the Walls of the Town of Nuremberg, Germany.*

It is a chalybeate water, with a fub-aftringent tafte, and contains alfo a faline matter.

It has been recommended in obftructions of the vifcera, and in female complaints.

## D'AX EN FOIX,

*About fifteen leagues weft of Thoulouse, France.*

This place abounds with hot fulphureous waters of different temperatures. They are recommended as baths, or otherwife, in thofe complaints in which the *Aix-la-Chapelle* and *Bareges* waters are ferviceable.

DED-

## DEDDINGTON,

This is a fulphureous chalybeate water; but foon lofes its fulphureous fmell by keeping.

Drank in large quantities it is purgative; and in lefs dofes as an alterative, it is good in fcorbutic and cutaneous diforders.

## DERBY,

*Near to the Town of Derby, in Derbyfhire.*

It is a chalybeate water of the nature of that of *Tunbridge*, but feems to be ftronger.

## DERRINDAFF,

*In the County of Cavan, Ireland.*

This is a fulphureous water, impregnated with a purging falt.

Its virtues refemble thofe of the *Afkeron* water.

## DERRYHENCE, or DERRYINCH,

*In the County of Fermanagh, Ireland.*

The water is fulphureous.

It contains alfo the foffil alkali, and re-fembles in its virtues the waters of *Drum-goon* and *Swadlingbar.*

K 2                    D E R.

## DERRYLESTER,

*In the County of Cavan, about three Miles from Swadlingbar, Ireland.*

The water is of the nature of that of *Swadlingbar*, but ſtronger.

## DOG and DUCK;

A noted tea-drinking houſe in St. George's Fields, near London; and in the ſpring and ſummer months the waters are much reſorted to.

The water is clear, and has but little taſte.

It is a mild purgative, and may be drank to the quantity of ſeveral pints. Moſt frequently, however, it is quickened by the addition of Glauber's, or other purging ſalts.

It is of uſe in ſcrophulous complaints, leproſies and diſeaſes of the ſkin; and is alſo ſaid to prevent the return of cancerous diſeaſes. For theſe complaints it may be uſed both internally and externally.

It

It is cooling and diuretic; and may be given freely to young people of robuſt conſtitutions. But it cools and relaxes people in years and of weak habits too much. It is alſo apt to bring on or encreaſe the fluor albus in weakly women.

## D O R T S H I L L,

*Near Litchfield, in Staffordſhire.*

The water is a briſk chalybeate, ſimilar to that of *Tunbridge.*

There is alſo a ſaline purging water of the nature of the *Barrowdale* water, but weaker.

## D R I G W E L L,

*Near Revenglas, in Cumberland.*

This is a briſk, ſpirituous, ſulphureous chalybeate; and in its virtues reſembles the *Deddington* water.

## D R O P P I N G W E L L,

*At Knareſborough, in Torkſhire.*

It is very cold, limpid, and ſweet taſted;

K 3 and

and in time petrifies fubftances thrown into it.

In its virtues it refembles the *Newton Dale* water. The dofe has formerly been feveral quarts in the day; but three or four half pints are now judged fufficient.

Its ufe fhould be preceded by a dofe or two of rhubarb.

## D R U M A S N A V E,

*Called likewife Mount Campbell, in the County of Leitrim, Ireland.*

This is one of the ftrongeft fulphureous waters in Ireland, as is fhewn by its quick and ftrong effect in difcolouring metals.

It is perfectly clear and limpid in common; but before rain becomes white.

It contains the foffil alkali, with a fmall quantity of purging falt.

It operates powerfully by urine, and purges fome conftitutions, but is faid to render others coftive.

D R U M=

# DRUMGOON,

## *In the County of* Fermanagh, Ireland.

The water has a ftrong fulphureous fmell, and tinges filver of a copper colour in a few minutes. It alfo depofits a black fediment at the bottom of the well.

It contains the foffil alkali, with a little fea falt.

It is recommended for the cure of cutaneous and fcrophulous diforders; and for worms.

There are two other fulphureous fprings in the neighbourhood; the one nearly refembles this; the other is more of a purgative nature.

# DUBLIN SALT SPRINGS.

*There are five of thofe Springs in* Francis Street, *and one in* Thomas's Court.

The waters are falt, and of the nature of *Barrowdale* water. For a purge, they muft be taken to the quantity of feveral pints. They operate without griping.

K 4                    DUL-

## D U L W I C H,

*Being is situated between Dulwich and Lewisham, in the County of Kent.*

The water is clear, and has a brackish taste, leaving a bitterness in the throat.

It contains a purging salt, together with sea salt.

This is a celebrated purging water; is also diuretic; and is recommended in a variety of diforders.

It is particularly of use in complaints arising from obstructions; as those of the liver, spleen, and other viscera.

It is recommended in the green sicknefs, the jaundice, the scurvy, in difficulty of urine, and in gravelly complaints.

It is said to strengthen the stomach, and create a good digestion.

It is also said to strengthen the nervous system, and therefore to be serviceable in palsies, apoplexies, and other nervous diforders.

orders. In thefe cafes it is beft taken warm.

The courfe of drinking this water is ufually twenty days. Three pints a day are to be drank at firft; it fhould be encreafed to eight pints by the tenth day, and afterwards decreafed in the fame manner.

A new fpring has fince been difcovered; whofe virtues are fimilar to thofe of the old one, but it is ftronger.

## D U N N A R D,

*About eighteen miles from Dublin.*

This is a chalybeate water, refembling that of *Peterhead,* but weaker.

## D U N S E,

*In Scotland.*

It is a chalybeate water, fimilar to that of *Tunbridge.*

DUR-

# D U R H A M.

*The spring is situated near Durham, on the north side of the river Ware.*

It is a ſtrong ſulphureous water, and is alſo impregnated with ſea ſalt.

In its virtues it reſembles the *Harrogate* water.

Near to this, in the middle of the river, is a ſalt ſpring, which is drank as a purging water.

# E G R A,

*In Bohemia.*

This is a ſpirity chalybeate water, and ope-rates both by ſtool and urine. It contains leſs fixed air than the *Pyrmont* water, but is more purgative, and ſeems rather to re-ſemble our *Scarborough* and *Cheltenham* waters.

# E P S O M,

*In Surry, about ſixteen miles from London.*

The water has a ſlight ſaline taſte, is clear, and without ſmell. But if it be kept

in

in covered veffels for fome weeks in the fummer it will ftink, and acquire a naufeous and faltifh bitter tafte.

This was the firft water from which the bitter purging falt (thence called *Epfom falt*) was obtained. But the falt ufually fold by that name is different from that yielded by the Epfom water, though perhaps not inferior in virtue. It is made from the bittern left after the chryftallization of common falt from fea water.

The Epfom water is purgative; for which purpofe it muft be drank to the quantity of two or three pints. It alfo operates by urine.

Taken in lefs quantity (about the third part of a pint three times a day) it is a mild alterative, and good in thofe complaints for which the *Acton* and *Pancras* waters are recommended.

It is likewife efteemed good for wafhing old fores.

F E L-

# FELSTEAD,

## *In Essex.*

The spring is situated at the bottom of a rock. The water is a light chalybeate, resembling that of *Islington*.

# FILAH,

## *Near Scarborough, in Yorkshire.*

This is a salt chalybeate water, and is used by the common people as a purgative; for which purpose they drink to the quantity of several quarts; it also operates by urine.

# FRANKFORT,

## *In Germany.*

There are two strong sulphureous waters in the neighbourhood of Frankfort on the Maine.

The one is called FAULPUMP,

The other FONS SCABIOSORUM.

They are also impregnated with sea salt,

and

and are of the nature of the *Moffat* and *Harrogate* waters.

## GAINSBOROUGH,

### *In Lincolnſhire.*

This is a weak ſulphureous chalybeate water, ſimilar to that of *Deddington.*

## GALWAY,

### *In the county of Galway, Ireland.*

It is a chalybeate water, of the nature of that of *Tunbridge.*

## GLANMILE,

### *Near Naul, in Ireland.*

It is a chalybeate water, reſembling that of *Peterhead,* but weaker.

## GLASTONBURY,

### *In Somerſetſhire.*

This water is of the ſame nature with thoſe of *Tilbury* and *Clifton;* but weaker than either of theſe.

It has alſo a ſmall mixture of ſea ſalt.

It

It is naturally fweet, but by keeping becomes putrid.

This water was formerly in great repute; and many fuperftitions were held concerning it; but it has not lately been efteemed.

Its virtues may be collected from what is faid of *Tilbury* water.

## G L E N D Y,

### *In the County of Mairns, Scotland.*

This is a ftrong chalybeate water, little inferior to that of *Peterhead.*

## G R A N S H A W,

### *Near Dunnaghadee, in the County of Down, Ireland.*

It is a chalybeate water, of the nature of that of *Caftle Connel.*

## G U G G A,

See *Kuka.*

H A I G H,

## H A I G H,

*Near Wigan, in Lancashire.*

It is impregnated with green vitriol; and is of the nature of the *Shadwell* water; which fee.

It works plentifully by vomit, and ftool; and is excellent for ftopping inward bleeding.

## H A M P S T E A D,

This is a chalybeate water, of the nature of that of *Islington,* but fomewhat ftronger. The dofe is from half a pint to feveral pints.

It was formerly, and perhaps defervedly, in great repute.

This water is better in the morning than in the middle of the day; and in cold, weather it is much ftronger than in hot.

## H A N B R I D G E,

*In Lancashire.*

It is a chalybeate water, of the nature of that of *Scarborough,* but lefs purgative.

H A N-

## H A N L Y S,

*Near Shrewſbury, in Shropſhire.*

The water is clear, and limpid, and has a ſaline and bitter, though not diſagreeable taſte.

It ſprings up with impetuoſity at the fountain; and does not change colour, or loſe its virtue, by being expoſed to the air.

It is purgative; and the doſe is from two to four half pints.

At this place there is alſo a *chalybeate water.* It is near to the purging water, and is of the nature of thoſe of *Scarborough* and *Landrindod.*

It is briſk and pungent to the taſte, and as it is taken from the fountain, clear, and not unpleaſant; but loſes its virtues by keeping.

## H A R R O G A T E,

*Near Knareſborough, Yorkſhire.*

There are four ſprings at this place, but the waters of all of them are nearly alike.

The

The water as it fprings up is clear and fparkling, and throws up a quantity of air bubbles.

It has a ftrong fmell of fulphur, and is fuppofed to be the ftrongeft fulphureous water in England.

It has a falt tafte, as it contains a confiderable quantity of fea falt, together with a little purging falt.

It is purgative; and the dofe required for this purpofe is about three or four pints.

When drank in fmaller quantities, it is a good alterative, and is found ferviceable in the fcurvy, king's evil, and difeafes of the fkin. It may be ufed at the fame time outwardly, by way of bath, or fomentation.

It has been found efficacious in deftroying worms.

It has been recommended in the gout, jaundice, the fpleen, the green ficknefs, and other diforders arifing from obftructions.

L                    It

It is ufed externally for removing old aches, ftrains, paralytic weakneffes, and the like. Alfo for the cure of ulcers, fcabs, the itch, &c.

N. B. Between *Harrogate* and *Knarefbo-rough*, are alfo feveral *chalybeate* waters, which feem to refemble thofe of *Hamp-ftead* and *Iflington*. The moft remarkable are, *the Allum Well, the Sweet Spaw,* and *the Tuewhet Well.* The latter is the ftrongeft.

## H A R T F E L L,

*In the county of Annandale, Scotland.*

It is impregnated with green vitriol, and refembles the *Shadwell* water.

It is recommended in inward bleedings, in immoderate flux of the menfes, in dyfenteries, in bloody urine, in the fluor albus, in gleets, in complaints of the ftomach and bowels, and in confumptions.

The dofe is from a gill to a pint or two, taken at repeated draughts in the morning.

Exter-

Externally, it cures itchy, and tetterous eruptions, and old fores, efpecially if taken at the fame time as an internal remedy.

# H A R T L E P O O L,

## *In Durham.*

This is a fine clear chalybeate water; with a flight fulphureous fmell, and pleafant tafte.

It is alfo diuretic and laxative; and is recommended as excellent in fcorbutic complaints, in bilious and nervous cholics, in pains of the ftomach and indigeftion, in the gravel, in female obftructions, in the hypo-chondriacal difeafe, in cachexy, in hectical heats, and in recent ulcers.

# H O L T,

## *Near Bradford, in Wiltfhire.*

The water is limpid, and has but little tafte.

It contains a purging falt, together with a large quantity of earth.

On

On account of the latter ingredient, it is but a very mild purge, and two quarts are ufually required to produce any confiderable operation this way.

Taken in lefs quantity it is alterative, and diuretic.

It is alfo good as a diluent, cooler, and ftrengthener; and creates an appetite.

Externally, rags, or fpunge dipt in it, are faid to cure fcrophulous ulcers, attended with carious bones; an internal courfe being obferved at the fame time.

It is alfo of fervice in old running ulcers of the legs, and other parts; in cutaneous foulneffes, though attended with hot corrofive humours; in the piles, in cancerous ulcers, and in foreneffes of the eyes. But in thefe cafes alfo it muft be ufed both internally and externally.

## H O L T,

See *Nevil-Holt.*

JES-

## J E S S O P's  W E L L,

*On Stoke Common, near Cobham, in Surry,*

This is a ſtrong purging water, with a nau-
ſeous taſte, and is alſo ſlightly chalybeate.

Drank to about a quart, it purges briſkly
without griping, and operates likewiſe by
urine.

It alſo enlivens the ſpirits, and as the doſe
is ſmaller than that of other purging waters,
it ſits better on the ſtomach.

It loſes its virtues by being kept.

Taken in leſs doſes as an alterative, it is
a good antiſcorbutic.

## I L M I N G T O N,

*In Warwickſhire, on the borders of Wor-
ceſterſhire.*

This is a very clear and ſparkling chalybeate water, abounding in fixed air, and
impregnated alſo with the foſſil alkali.

L 3                                    It

It preferves its virtues for feveral weeks in bottles well corked; though if expofed to the air, it lofes them in twenty-four hours.

It operates by urine; though it alfo fometimes purges.

It is recommended as excellent in fcorbutic complaints, in obftructions of the liver and fpleen, in the jaundice, in beginning dropfies, in the gravel, and obftruction of urine, and in diforders arifing from acidity.

Externally, it is good for old running fores, fcorbutic eruptions, and the like.

## I N G L E W H I T E,

### *In Lancafhire,*

It is a ftrong chalybeate fulphureous water, and is good in fcorbutic, and cutaneous difeafes. But it will not purge unlefs Glauber's, or other falt be added to it.

## I S L I N G T O N,

### *In the county of Middlefex, near London,*

This is a light chalybeate water, ftriking a purple or blackifh colour with galls, and is
reckoned

reckoned one of the beft of the kind about London.

The iron in this water is held in folu-tioŋ by means of *fixed air*, or *aërial acid*, as in the Pyrmont water. If after the fixed air has efcaped, and the iron (which it held in folution) precipitates, the water be left to putrify, the fixed air difengaged by the putrefaction again diffolves the iron, and caufes it to be fufpended in the water; it then recovers its chalybeate tafte, and property of tinging with galls, both which it had loft before.

It is recommended in indigeftion, and lofs of appetite, in lownefs of fpirits, nervous, hyfteric, and hypochondriacal complaints, and relaxed conftitutions, and raifes the fpirits greatly. It is good in the fluor albus, in weakneffes from mifcarriage, in obftruc-tions of the liver, the kidnies, &c. It is al-fo ferviceable in difeafes of the fkin, in fcor-butic complaints, in the gravel, and in pa-ralytic diforders.

It operates chiefly by urine, and may be drank to the quantity of feveral half pints, or

even pints, according to the patient's conftitution.

This water was formerly in great repute,
and deferves to be more frequented than it is
at prefent.

## K A N T U R K,

*In the county of Cork, Ireland.*

It is a chalybeate water, of the nature of
that of *Peterhead,* but weaker.

## K E D D L E S T O N E,

*In Derbyſhire.*

This is a ftrong fulphureous water, and
ftinks intolerably.

It is extremely clear at the fountain, but
by ftanding becomes blackifh. It prefently
turns filver of a black copper colour.

It is impregnated with fea falt.

Its virtues refemble thofe of the *Harro-*
*gate* water.

# KENSINGTON,

*In the county of Middlesex, near London.*

It is a purging water, of the nature of those of *Acton* and *Pancras*.

# KILBREW,

*In the county of Meath, Ireland.*

This is a strong vitriolic chalybeate water, and resembles the *Shadwell* water.

Half a pint vomits and purges.

When taken as an alterative it should be used with great caution, beginning with a small quantity, and increasing the dose.

It is recommended in the fluor albus, in immoderate fluxes from the womb, in obstinate intermittents and in dropsies.

# KILBURN,

*In Middlesex, near London.*

It is a purging water, like those of *Bagnigge Wells, Dulwich,* &c.

KIL-

## K·I L R·O O T,

*In the county of Antrim, Ireland.*

It is of the nature of *Barrowdale* water, but weaker; feveral pints being required for a purge.

## K I L L I N G S H A N V·A L L Y,

. *In the county of Fermanagh, Ireland,*

This is·a chalybeate water, and is alfo diuretic and gently laxative. It feems to refemble the *Hanlys* chalybeate water.

## K I L L A S H E R,

*In the county of Fermanagh, Ireland.* ·

The water is ftrongly fulphureous, and it contains the foffil alkali.

Its virtues. refemble thofe of teh *Swadlingbar* water.

## K I N A L T O N, or K Y N O L T O N,

*A village in Nottinghamfhire.*

The water is limpid and cooling, with a fomewhat faltifh tafte.

It

It is purging; but is weaker than the Epfom water, and therefore muſt be drank plentifully.

## KINCARDINE,

*In the county of Mairns, Scotland.*

This is a chalybeate water, little inferior in ſtrength to that of *Peterhead.*

## KINGSCLIFF,

*In Northamptonſhire.*

It is a chalybeate laxative water, and re-fembles the *Scarborough* and *Cheltenham* waters.

## KIRBY, or KIRKBY-THOWER,

*In Weſtmoreland.*

There are two ſprings nearly of the fame kind, only the lower one is reckoned the ſtrongeſt chalybeate.

The water of both is clear, fine, and has a chalybeate fweetiſh taſte. Drank to the quantity

quantity of feveral quarts it is purgative. It is alfo a good corrector of acidities.

## KNARESBOROUGH,

See *Dropping Well.*

## KNOWSLEY,

### *In Lancafhire.*

This is a light fpirituous chalybeate water, and both taftes and fmells of iron.

If drank to four or five pints it is laxative.

It refembles the *Scarborough* and *Cheltenham* waters in virtue, though it feems to be lefs purgative.

## KORYTNA,

### *Near Hunnobroda, in Moravia, Germany.*

It is fituated on an high and almoft inacceffible rock, in the midft of a thick wood.

It has a very fœtid difagreeable tafte, and a black colour; and there is much mud at the bottom of the well.

It

It is reckoned the ſtrongeſt ſulphureous water in that country.

## K U K A,

*In the county of Graditz, in Bohemia, near the town of Jaromitz, at the conflux of the rivers Elbe and Orlitz, Germany.*

This is a very briſk chalybeate water, highly impregnated with fixed air, and alſo with the foſſil alkali. It has a grateful and ſomewhat pleaſant taſte, and a pungent ſmell, which affects the whole head. If it be heated, it emits a penetrating acid ſulphureous ſmelling vapour. It will not bear carriage.

It operates chiefly by inſenſible perſpiration; and ſometimes by ſpitting, by ſweat, and by urine.

In its general virtues it reſembles the *German Spaw* waters.

## LA MARQUISE, et LA MARIE.
See *Vahls.*

LAN-

## LANCASTER, or SALE's SPA,

*In Lancashire.*

This is a clear chalybeate water, power-fully diuretic, gently purgative, and vomits if taken to the quantity of several quarts.

Taken as an alterative it has the general virtues of the *Tunbridge* water.

## L A T H A M,

*In Lancashire.*

It is a fine chryftalline chalybeate, of the nature of the *Tunbridge* water.

## L L A N D R I N D O D,

*In the county of Radnor, South Wales.*

In this place there are three mineral springs :

1ft. *The faline pump, or purging water.*

2d. *The fulphureous water,* commonly called the *black ftinking well.*

3. *The chalybeate rock water.*

The

The *faline purging*, or *pump water*, may be ufed as a *purge* twice in a week. It is directed to be drank at the fountain head by half pints, till it begins to operate; the patient walking or riding about between each draught. It operates alfo by urine.

For an *alterative*, about three pints are directed to be drank in a day. A pint and half in the morning before breakfaft, at three draughts, a quarter of an hour between each half pint. The other pint and half likewife at three draughts : one an hour before dinner; another about fix o'clock in the evening; and the third going to bed. If the body remains coftive, the quantity muft be increafed. The courfe fhould be continued feveral weeks; and the moft proper feafon is the fummer.

It is alfo ufed as a bath and fomentation.

It is recommended both internally and externally in the fcurvy, leprofy, tetters, King's evil, and all cutaneous foulneffes.

It is alfo prefcribed in the gravel, the
hypo-

hypochondriacal difeafe, indigeftion, and in other complaints.

*The fulphureous water;* called alfo the *black ftinking water,* from its ftrong fmell, and the blacknefs of the channel through which it paffes.

The quantity to be drank cannot in general be determined. But it is beft to begin with fmall dofes, from a pint to a quart in the morning, taken at repeated draughts. The quantity may be encreafed as the conftitution will bear; or as much as will fet eafy on the ftomach, and pafs off well. When it gives the leaft uneafinefs, it is a fign that the dofe is too large.

It is alfo ufed outwardly, by way of bath or fomentation.

It is recommended in a variety of complaints. In the King's evil, fcurvy, leprofy, and all cutaneous difeafes. In the jaundice, hypochondriacal, and other diforders arifing from obftruction. In the gravel, rheumatifm, gout, bloody flux,

heclic

hectic fever, weakneffes of the limbs, want of digeftion, and many others.

The *chalybeate,* or *rock water,* is limpid and tranfparent, as taken from the fountain, but on ftanding foon lofes thefe qualities, together with its chalybeate tafte. Mixed with fugar and rough cyder as it is taken up from the fpring, it excites a brifk fermentation.

It is recommended in fuch chronic diftempers as proceed from laxity of the fibres, and weaknefs of the mufcular fyftem. In weaknefs of the nerves; in paralytic complaints; and the like.

It is alfo good in fcorbutic cafes; in moift and convulfive afthmas; in obftinate agues; in obftructions of the lower belly; in wandering, flow, nervous fevers; and in diforders arifing from obftruction.

## L L A N G Y B I,

*In Caernarvonfhire, North Wales.*

The water has an harfh tafte, inclining to bitter.

M                    It

It has been found efficacious in diforders of the eyes; in the King's evil; fcald heads; ulcers; eruptions of the fkin; the fcurvy; the itch, &c. Alfo in rheumatifms, palfy, and convulfions fits.

This water has long been in repute in the neighbourhood.

## LEAMINGTON.

This is of the nature of *Barrowdale water,* but much weaker. The dofe for a purge is from a quart to four or five pints, and it likewife ufually vomits.

## LEEZ,

### *Near the Earl of Manchefter's, Effex.*

It is a chalybeate water, fimilar to thofe of *Iflington* and *Hampftead.*

## LINCOMB,

### *Near Bath, in Somerfetfhire.*

This is a chalybeate and acidulous water, containing alfo the foffil alkali, with a fmall quantity of purging falt. It foon lofes its

virtue

virtue if expofed to the air; and in a few days alfo in bottles.

It refembles, in its virtues, the waters of *Thetford* and *Ilmington.*

## L I S B E A K,

*In the parifh of Killafher, in the county of Fermanagh, Ireland.*

Here are two ftrong fulphureous waters, much of the fame kind.

Their contents and virtues refemble thofe of the *Swadlingbar* water.

## L I S-D O N E-V A R N A,

*In the county of Clare, in Ireland.*

This is a ftrong chalybeate water, and manifefts itfelf as fuch both to the tafte and fmell. It is alfo impregnated with the foffil alkali.

It keeps its virtue in well-corked bottles.

It ufually vomits and purges on firft drinking, but afterwards operates by urine.

M 2                                   It

It feems to refemble, in virtues, the *Thetford* and *Ilmington* waters.

## LOANSBURY,

*In Lord Burlington's park, in Yorkſhire.*

This is a fulphureous water, weakly impregnated with a purging falt.

It feems to be of the nature of the *Aſkeron* water; but is only uſed at prefent for waſhing mangy dogs and fcabby horfes.

## MACCROOMP,

*In Ireland, about ſixteen miles from Cork.*

This is a chalybeate water, impregnated with the foſſil alkali, and refembles the *Thetford* and *Ilmington* waters.

## MAHEREBERG,

*Situated near Brenden Bay, in the county of Kerry, Ireland.*

It is of the nature of the *Barrowdale* water. The dofe for a purge is from a pint to a quart.

MALLOW,

# M A L L O W,

## *In the county of Cork, Ireland.*

This is a warm water, perfectly limpid and pleasant tasted, and keeps long in bottles well corked.

It is recommended in most cases for which the *Bristol* water has been used.

# M A L T O N,

## *The spring lies at the west end of the town of New Malton, in Yorkshire.*

It is a strong chalybeate, abounding with fixed air when fresh drawn; has a saltish taste, and leaves a bitterness in the throat.

It operates by stool and urine. The dose is from three pints to twice that quantity. If the stomach be foul, it is apt to vomit. In its virtues it resembles the *Scarborough* water.

# M A L V E R N,

## *In Gloucestershire.*

There are two noted springs at this place,

M 3                                                    one

one of them called the *Holy Well*, in the midway between Great and Little Malvern, the other is about a quarter of a mile from Great Malvern. But the waters are not materially different.

They are light and pleasant chalybeates, and are remarkable for being almost entirely free from any earthy matter. For three quarts of the Holy Well water being evaporated, scarce the fourth part of a grain of sediment was left behind.

They are recommended as excellent in diseases of the skin; in leprosies; scorbutic complaints; the King's evil; glandular obstructions; scald heads; old sores; cancers, &c. They are also serviceable in inflammations and other diseases of the eyes; in the gout and stone; in cachectic, bilious, and paralytic cases; in old head achs, and in female obstructions.

The external use is by washing the part under the spout several times in a day; afterwards covering the part with cloths dipt in the water, which must be kept

constantly

conftantly moift. Thofe who bathe, ufually, go into the water with their linen on, and drefs upon it wet, and it is never found to be attended with inconvenience.

The waters, when firft drank, are apt to occafion, in fome, a flight naufea; others they purge brifkly for feveral days; but they operate by urine in all.

It is advifeable to drink freely of the waters for fome days before they are ufed externally.

## M A R K S H A L L,
### *In Effex.*

This is a chalybeate water, refembling thofe of *Iflington* and *Hamftead.*

## M A T L O C K,
### *Near Wirkfworth, in Derbyfhire.*

At this place (which is perfectly romantic) are feveral fprings of warm water, which appear to be of the nature of the *Briftol* water, except that it is very flightly impregnated with iron.

M 4        Its

Its virtues are fimilar to thofe of the *Briftol* and *Buxton* waters.

The baths are recommended in rheumatic complaints, in cutaneous diforders, and in other cafes where warm bathing is ferviceable.

There are great numbers of petrifactions in the courfe of this water.

## MAUDSLEY,

*Near Prefton, in Lancafhire.*

The water is of a bluifh colour, has a fœtid fmell, and a brackifh tafte.

It is a ftrong fulphureous water, and is alfo impregnated with fea falt.

It is purgative, and has nearly the fame virtues as the *Harrogate* water.

## MECHAN,

*In the county of Fermanagh, Ireland.*

In this place there are two fulphureous fprings, both of the fame nature.

They

They contain the foffil alkali, and in their virtues refemble the *Drumgoon* and *Swad-. lingbar* waters.

## M I L L A R's  S P A W,

*Stockport, in the county of Lancaſter.*

It is a chalybeate water of the nature of that of *Tunbridge*, and feems to be ftronger of the iron.

## M  O  F  F  A  T,

*In the county of Annandale, Scotland.*

At this place there are two fprings or wells.

They are both fulphureous, and have a ftrong fmell and tafte; the upper one is the ftrongeft, and moft naufeous, and is lefs drank of than the other, though as it bears heat better it is moft ufed for bathing.

The Moffat water is alterative, and diuretic, but it fometimes purges.

The quantity to be drank is three or four Englifh quarts in a day. Thofe of weakly conftituconſtitutions, and children, leſs in proportion.

It is recommended to clear the firſt paſſages previous to a courſe of the water; and ſalts are alſo directed to be diſſolved in it if it does not paſs off readily. Letting blood is adviſed if it ſtays in the body too long.

The water muſt be refrained from if the patient has a cough, and in tubercles of the lungs. In hectic caſes it muſt alſo be drank with caution. Theſe caſes excepted, it may be uſed with great freedom.

It is much recommended in diſorders of the ſkin, in ſcrophulous and ſcorbutic complaints, in chronic obſtructions, in female weakneſſes and barrenneſs, in cholics, and other pains of the ſtomach and bowels.

Externally, it is employed for waſhing ſcrophulous ſores, and foul ulcers; cloths wetted with it are alſo directed to be applied. It has been warmed and uſed by way of bath

bath to particular parts, and to the whole body; but it fhould not be made too hot.

The hot fteam has been ufed with fuccefs for foftening and relaxing hard fwelled parts, and ftiff joints.

In thefe and the like complaints, it may be ufed both internally and externally.

## M O S S   H O U S E,
### *Near Maudfley, in Lancafhire.*

This is a brifk chalybeate water, and in its virtues refembles thofe of *Hampftead* and *Iflington*.

## MORETON, or MORETON-SEE.
*Situated about two miles Weft of Market-Drayton, in Shropfhire.*

It is efteemed as an excellent cooling and diuretic purge. It operates brifkly; is pungent to the tafte, and feems to be of the nature of *Holt* water.

MOUNT

# M O U N T   D' O R,

*Seven leagues from Clermont, in the Auvergne,*
*France.*

This water is warm, and of the nature of
the *Aix-la-Chapelle*.

Taken internally it acts as a diuretic,
and it sometimes purges. Bathing in it
sweats profusely, without weakening the
patient.

It has been recommended in the rheuma-
tism, gout, palsy, and many other diforders.

# M O U N T     P A L L A S,

*In the county of Cavan, Ireland.*

It is a chalybeate water, and seems to be
of the nature of the *Athlone*.

# N E V I L - H O L T,

*Near Market-Harborough, in Leiceſterſhire.*

This is an exceeding clear water as it falls
from the spout, and is void of all smell.

It

It has a brifk auftere bitter, yet not difagreeable tafte, and abounds in fixed air.

Expofed to the air, it foon becomes turbid, and fpoils. But in well clofed bottles
it will keep good.

Drank to the quantity of feveral pints,
it proves purgative, and operates without
griping.

It alfo operates by urine and fweat.

It is a powerful antifeptic in putrid difeafes.

When taken as an alterative, it muft be
taken in fmall dofes, from a few fpoonfuls,
to a quarter or half a pint feveral times in a
day, according to its effect; and a little
brandy, or the like, may be added if it fits
cold on the ftomach.

It is efteemed an excellent remedy in old
dyfenteries and diarrhœas, in internal he-
morrhages, in the fluor albus, and gleets,
in the gravel, in rheumatifms, and for the
worms ;

worms; it is good in atrophies, in bloated conftitutions, and dropfical complaints, in fcorbutic diforders, in want of appetite, and in other cafes; in inflammatory complaints however, and where there is an acidity of the humours, it does mifchief.

Externally, it is a fpeedy cure for frefh wounds, for inflamed eyes, and hectic ulcers, &c. efpecially if taken inwardly at the fame time.

## NEW CARTMAL,

See *Rougham.*

## NEWNHAM REGIS,

*In Warwickfhire.*

There are three wells at this place: they are all of them chalybeate, laxative, and diuretic; and feem to refemble the *Scarborough* water.

They have fomewhat of a fulphureous fmell.

NEW-

# NEWTON DALE,

*In the North Riding of Yorkſhire.*

This is a cold petrifying water.

It is ſaid to cure effectually, loofeneſſes, and bleedings of every kind; and that it reſtores weakened joints, though beginning to be diſtorted, by bathing in it,

# NEWTON STEWART,

*Near Caſtlehill, in the county of Tyrone, Ireland.*

It is a chalybeate water, of the nature of that of *Tunbridge.*

# NEZDENICE,

*In Germany.*

*About half a mile from Hunnobroda, in the diſtrict of the Caſtle of Banow. The ſpring is near this village.*

This is an acidulous water, impregnated with fixed air like thoſe of *Seltzer* and *Pyrmont.*

It

It is in great repute among the people in the neighbourhood, for the cure of many diforders, particularly thofe in which the waters juſt mentioned are ferviceable.

## N O B B E R,

### *In the county of Meath, Ireland.*

It is a vitriolic water, and refembles thofe of *Hartfell* and *Crofstown.*

## N O R M A N B Y,

### *Four miles from Pickering, in Yorkſhire.*

It is clear, beautiful, and fœtid, and when poured out fparkles like Champagne.

It is a fulphureous, and gently purgative water, and refembles the *Aſkeron* water in virtues.

## N O R T H-H A L L,

See *Barnet.*

N O T-

# N O T T I N G T O N,

*Near Weymouth, in Dorfetfhire.*

This is a ftrong fulphureous water.

Its flavour refembles that of boiled eg3s; and its colour, in a tin veffel, is blue. A fhilling put into it at the fountain head, becomes, in a few minutes, blue.

It contains the foffil alkali, with a little earth.

It is in repute for curing foulneffes of the fkinr

# O R S T O N,

*In the county of Nottingham, near Thoroton.*

This water has a delicious, gentle, rough, fweetifh, chalybeate tafte, and a flight fulphureous mell. It is replete with fixed air, fparkles and flies when poured out into a glafs, and makes the heads of thofe who drink it giddy.

It foon fpoils by expofure to air.

It is purgative, and feems to be poffeffed of the fame virtues as the *Hartlepool* water. It makes the tongue and ftools black, like many other chalybeate waters.

N                O U L-

# O U L T O N,

*In Norfolk.*

It is a light chalybeate water, fimilar to that of *Iflington.*

# O W E N   B R E U N,

*In the county of Cavan, Ireland.*

This is a fulphureous watèr, impregnated with a purging falt, and a little native al-kali.

Its virtues refemble thofe of the *Afkeron* water.

# P A N C R A S,

*In Middlefex, near London.*

The water is almoft infipid to the tafte.

It is impregnated with a purging falt, to-gether with a fmall portion of fea falt.

It is therefore a purgative water, and is alfo diuretic.

Its virtues are allied to thofe of the *Chel-tenham*

*tenham* water, and it is alfo of fervice in the ftone, gravel, and fimilar diforders.

## P  A  S  S  Y,

*Near Paris, in France.*

It is a clear, colourlefs, chalybeate water, with a fubacid tafte, and ferruginous fmell, and emits plenty of air bubbles.

It is a ftrong chalybeate water, but does not prove purgative, unlefs drank in large quantity. It is of the nature of *Pyrmont* water.

## P E T E R H E A D,

*In the county of Aberdeen, Scotland.*

This is one of the ftrongeft, and moft famous chalybeate waters in Scotland. It is of the nature of our *Iflington* water, but more powerful.

## P E T T I G O E,

*In the county of Donnegal, Ireland.*

It is one of the ftrongeft fulphur waters

in

in Ireland; and is alſo impregnated with a purging ſalt.

In its virtues it reſembles the *Aſkeron* water.

## P L O M B I E R S,

*In Loraine, France.*

The water is tepid and ſaponaceous, with a ſaltiſh taſte.

It is recommended for complaints of the ſtomach proceeding from acidity; in ſpitting of blood; in hæmorrhages; phthyſical and aſthmatic complaints; in dropſy of the belly; the diabetes; fluor albus; dyſentery; and in all cutaneous diſorders.

It is drank from a pint to three quarts, on an empty ſtomach, in the morning; it is diuretic and laxative.

It is alſo uſed outwardly as a bath; and is reckoned excellent for waſhing ulcers.

## P O N T G I B A U L T,

*In the Auvergne, France.*

The water is limpid, ſubacid, and vinous.

It

It contains the native alkali, together with calcareous earth.

It is diuretic and gently opening; and its virtues are allied to thofe of the *Tilbury* and *Seltzer* waters.

## P Y R M O N T,

*In Weftphalia, Germany.*

This is a very brifk, fpirity chalybeate, abounding in fixed air; and when taken up from the fountain, fparkles like the brifkeft Champaign wine. It has a fine, pleafant, vinous tafte, and a fomewhat fulphureous fmell. It is perfectly clear, and bears carriage better than the *Spaw* water.

Perfons who drink it at the well are affected with a kind of giddinefs or intoxication; owing, it may be fuppofed, to the great quantity of fixed air with which the water abounds.

The common operation of this water is by urine; but it is alfo gently fudorific; and if taken in large quantity proves

N 3

laxative,

laxative. When, however, it is required to have this latter effect, it is ufual to mix fome falts with the firft glaffes.

It is drank by glafsfuls in the morning to the quantity of from one, to five or fix pints, according to circumftances, walking about between each glafs.

It is recommended in cafes where the conftitution is relaxed. In want of appetite and digeftion ; in weaknefs of the ftomach, and in heart-burn ; in the green ficknefs ; in female obftructions, and in barrennefs ; in the fcurvy, and cutaneous difeafes ; and in the gout, efpecially when mixed with milk. In cholics ; in bloody fluxes ; in difeafes of the breaft and lungs ; in which cafes it is beft taken lukewarm ; in nervous, hyfteric, and hypochondriacal diforders ; in apoplexies and palfies ; in the gravel, and urinary obftructions ; in foulnefs of the blood ; and in obftructions of the finer veffels. It amends the lax texture of the blood ; exhilarates the fpirits without inflaming, as vinous liquors are apt to do ;

and

and is among the beft reftoratives in de-
cayed and broken conftitutions.

This water poffeffes the general virtues
of the *Spaw* water ; and at the fountain it
is even more fpirity, as well as a ftronger
chalybeate. The reader therefore is referred to what is faid of the *Spaw* water,
for a further account of its virtues.

## QUEEN CAMEL,

*Near Sherborne, in Somerfetfhire.*

The water has a fœtid, fulphureous fmell,
like the wafhings of a foul gun. It tinges
filver of a yellow or black colour, and
blackens the ftones on which it runs. It
is alfo faid to be colder than common
water.

It contains the foffil alkali, together with
fea falt, a chalky earth, and a bituminous
or fulphureous matter.

It has been ufed with fuccefs both inwardly and outwardly in cutaneous diforders, the fcurvy, and the King's evil.

N 4 And

And for thefe purpofes a place is contrived for bathing.

## R I C H M O N D,

*In the county of Surry.*

This is a purging water, of the nature of thofe of *Acton* and *Pancras.*

## R O A D,

*In Wiltfhire.*

This a chalybeate water with a fulphu‑reous fmell, and is impregnated with the foffil alkali.

It is recommended internally and ex‑ternally in fcorbutic and fcrophulous cafes, and in cutaneous difeafes, &c.

On firft taking this water it acts as a gentle purge.

It does not bear carriage.

## R O U G H A M,

*In Lancafhire.*

It is of the nature of *Barrowdale* water, but much weaker.

The

The dofe for a purge is from three to eight quarts. In that quantity it operates powerfully by ftool, and alfo by urine.

## SAINT BARTHOLOMEW's WELL, *Ireland.*

*It is about two miles fouth-weft from Cork.*

The water is foft, and mixes fmoothly with foap.

By keeping it putrifies, and then tinges filver, and throws up a ftinking fcum which has fomewhat of an irony tafte. Galls then give it a purple tinge, which they do not to the frefh water.

It contains an alkaline falt.

Its virtues are fimilar to thofe of the *Tilbury* water.

## SAINT ERASMUS's WELL.

*Situated on Lord Chetwynd's grounds in Staffordfhire.*

The water is of the nature of *Barrow-dale*, but much weaker.

It

It is of the colour of fack, but without much tafte or fmell.

If drank to the quantity of feveral quarts, it operates powerfully by ftool.

## SALES SPAW.

See *Lancafter.*

## SCARBOROUGH,

### *In Yorkfhire.*

The waters of this place are chalybeate and purging; and they are more frequented and ufed than any other water of this clafs in England.

There are two wells; the one more pur-gative, the other a ftronger chalybeate. Hence the latter (which is neareft the town) has been called the *chalybeate* fpring, the other the *purging*; though they are both impregnated with the fame principles, but in different proportions. The *purging* is the moft famed, and is that which is ufually called the *Scarborough* water.

When

When thefe waters are poured out of one glafs into another, they throw up a number of air bubbles; and if fhook for a while in a clofe ftopt phial, and the phial be fuddenly opened before the commotion ceafes, they difplode an elaftic vapour with an audible noife, which fhews that they abound in fixed air.

At the fountain they both have a brifk, pungent, chalybeate tafte; but the *purging* water taftes bitterifh, which is not ufually the cafe with the *chalybeate* one.

They lofe their chalybeate virtues by ex-pofure, and alfo by keeping; but the purging water fooneft.

They both putrify by keeping; but in time recover their fweetnefs.

Four or five half pints of the *purging* water drank within an hour, give two or three eafy motions, and raife the fpirits. The like quantity of the *chalybeate* purges lefs, but exhilarates more, and paffes off chiefly by urine.

Thefe

Thefe waters have been found of fervice in hectic fevers, in weakneffes of the ftomach, and indigeftion ; in relaxations of the fyftem ; in nervous, hyfteric, and hypochondriacal diforders ; in the green ficknefs, in the fcurvy, rheumatifm, and afthmatic complaints ; in gleets, the fluor albus, and other preternatural evacuations, and in habitual coftivenefs. The waters are to be varied according to the intention to be anfwered.

## S C O L L I E N S E S,

### *In Upper Rhoetia, Switzerland.*

It is a chalybeate water, impregnated with the foffil alkali ; and fo full of fixed air, that it often burfts the bottles in which it is kept.

It makes the drinkers giddy, and operates mildly, though largely by ftool, and by fpitting.

It is efteemed excellent for cholicky pains, both as a cure and preventative.

In its general virtues it refembles the *Spaw* water.

S E A

# SEA WATER,

Sea water has a falt, bitterifh tafte, appears of a greenifh colour, and becomes fœtid by keeping.

As an immenfe number of fprings, rivers, &c. are continually emptying themfelves in the fea, as it contains an almoft infinity of animals and vegetables, and covers and wafhes fuch a variety of lands and fhores, it would feem to be impregnated with very heterogeneous matters. Neverthelefs, the water, in different parts of the ocean, appears to be nearly alike, and the difference in its contents appears to be much lefs than might at firft be imagined.

It contains fea falt in great quantity. A chalky earth, or rather, what is called by the chemifts, a calcareous marine falt: a bittern, from which the Epfom falts of the fhops are prepared, and fome unctuous and bituminous matters.

Sea water, in hotter climates, generally contains a greater proportion of thefe matters than that in colder ones, and therefore

is

is ftronger. The difference, in fome places, is above two to one.

Sea water taken internally, in fmall quantity, proves a ftimulating and heating remedy, diffipating the finer fluids, and occafioning thirft.

In larger quantity it proves purgative. But differs from other purges in that patients who drink it daily for a confiderable time, inftead of lofing, often gain ftrength by it.

It therefore acts, not merely as a purgative, but gives alfo a brifk ftimulus to the ftomach and inteftines, thereby encreafing the appetite, and promoting digeftion.

By means of this excellent property of fea water (viz. our being able to keep up a purging for a confiderable time, without hurting the conftitution), we are enabled frequently to remove diforders which have refifted the force of other remedies.

It is of excellent ufe in fcrophulous complaints ;

plaints ; and glandular fwellings are generally
removed by it. If joined with the bark, it
has fometimes a better effect in thofe cafes.

It is alfo ferviceable in purging off grofs
humours, which have been the confequence
of indulging the appetite too freely, and
leading an inactive life : alfo in cleanfing the
inteftines of vifcid mucus, and worms.

In cafes where there is fever, heat, or
inflammation, fea water is found to be hurtful. Previous to its ufe, therefore, thefe
fymptoms fhould be removed by bleeding,
purging, and a proper cooling treatment.

As Sea water is fpecifically heavier than
common water, and, (by reafon of the faline matters contained in it,) is alfo more
ftimulating, it is more efficacious when ufed
externally as a bath.

It has alfo particular virtues when externally ufed. On account of its ftimulating
and difcutient property, it is excellent in the
fcrophula or king's evil, in hard fwellings,
in the bite of a mad dog, in the rickets, in
the

the dry leprofy, and itch, in paralytic and fcorbutic complaints, and in many other cafes. But in moft of thefe, it is proper to ufe it both internally and externally.

## S   E   D   L   I   T   Z,

*Germany,*

*A village in Bohemia.*

This purging water is of the fame nature as our Epfom, but much ftronger.

Two or three tea cups full are generally fufficient for a dofe; and the ftrongeft conftitution fcarce requires more than a pint.

## S   E   L   T   Z   E   R,

*In Germany.*

*This fpring is near to the town of Neider, or Lower Seltzer, about three leagues from Franckfort on the Maine, in the Lower Archbifhoprick of Treves.*

It rifes near a fmall trout ftream. The country and avenues around are delightful, and afford a very pleafing profpect.

The

The water iffues forth with great rapidity; is remarkably clear and light, and on pouring it from one veffel to another, plenty of air bubbles arife.

It has, at firft, fomewhat of a brifk fub-acid pungent tafte, but leaves behind a lixivial one.

· If expofed twenty-four hours to the air, it lofes entirely its original tafte, and acquires that of a flat alkaline ley, But no fediment is depofited.

It putrifies fooner than any other medicinal water.

When frefh, it makes an immediate effervefcence with acids, but efpecially with Rhenifh wines, and a little powdered fugar,

It alfo curdles with a folution of foap,

It does not change purple with galls; and therefore contains no chalybeate.

O                    Oil

Oil of tartar dropt into it makes it milky, but does not occafion a precipitate.

It contains an alkaline falt, together with a fmall quantity of calcareous earth, and fea falt, and it abounds in fixed air.

The operation of this water is chiefly by urine, for it has no purgative virtues. It corrects acidities, renders the blood and juices more fluid, and promotes a brifk and free circulation. Hence it is good in ob- ftructions of the glands, and againft grofs, and vifcid humours.

It is of great ufe in the gravel and ftone, and in other diforders of the kidnies and bladder.

It is alfo excellent in gouty and rheuma- tic complaints *, efpecially when mixed with milk.

† It is drank with great fuccefs in fcorbu- tic, cutaneous, and putrid diforders.

---

* In thefe diforders its virtue is faid to be much im- proved by the addition of Rhenifh wine, and a little fugar.

† Affes, or goats milk, is ufually preferred.

It

It is good againſt the heart-burn; and it is alſo an excellent ſtomachic. Several pints may be drank in the courſe of a day.

On account of its diuretic quality, it is of ſervice in dropſical complaints.

Mixed with aſſes milk, it is of great uſe in conſumptive caſes, and in diſorders of the lungs.

It is in great eſteem in nervous diſorders, either with, or without milk, as is found to be moſt ſuitable to the conſtitution.

It is alſo of ſervice in hypochondriacal, and hyſteric complaints, and in obſtructions of the menſes, eſpecially if exerciſe be uſed.

It is given in purgings and fluxes ariſing from acidity in the bowels, with good ſucceſs.

Drank by nurſes, it is ſaid to render their milk more wholeſome and nouriſhing to children, and to prevent it from turning four on their ſtomachs.

O 2　　　As

As the fixed air of this water fo foon flies off, it ought either to be drank on the fpot, or at leaft fhould be impregnated with a frefh quantity previous to its being taken, according to the directions given in the beginning of this treatife.

Thofe perfons, with whofe ftomachs water alone does not fo well agree, are advifed to mix with it fome generous and agreeable wine, in cafes where wine will not be hurtful. (See alfo *Spaw* and *Pyrmont* waters.)

## S E N E, or  S E N D,

*Near the Devizes, Wiltfhire.*

At this plaçe are two chalybeate fprings, one of them ftronger than the other, but both refembling in virtues the *Hampftead* and *Iflington* waters.

They are diuretic, but not purgative.

At a village called *Paulfholt*, near this place, is another chalybeate fpring.

S E Y D-

## S E Y D S C H U T Z,

*In Germany.*

It is fituated near to that of *Sedlitz*, and is of the fame purgative nature, but fomewhat ftronger.

## S H A D W E L L,

*Near London, fituated in Sun Tavern Fields.*

This is a vitriolic chalybeate water; that is, it is impregnated with green copperas, or vitriol of iron, and is one of the ftrongeft waters of the kind in England; it alfo contains iron held in folution by fixed air, or aërial acid.

It has an acid, auftere, vitriolic tafte, and with galls, gives a bluifh black colour like ink.

It has been taken inwardly to the quantity of a pint, divided into two or three dofes in the courfe of an hour in the morning. It vomits, and gently purges; it turns the ftools black.

O 3                    It

It has been found of fervice in the fluor albus, in obftinate gleets, and dyfenteries; in inward bleedings; in the jaundice; and in fcorbutic, and leprous cafes. But it has chiefly been ufed externally for fore eyes, the itch, fcabs, tetters, fcald-head, ulcers, fiftu-las, and the like, by wafhing, or elfe applying linen rags dipped in it, to the parts.

In fcorbutic and leprous cafes, the inter-nal ufe is firft advifed 'till the eruptions are thrown out, which are then to be removed by the outward application of the water.

## S H A P M O O R,

*The fpring is fituated in a marfhy heath, be-tween Shap and Orton, in Weftmoreland.*

This is a fulphureous water, impregnated with a purging falt.

Three pints of it prove purgative.

In its virtues it feems to refemble the *Afkeron* water.

SHAT-

## S H E T T L E W O O D,

*Situated between Bolfover and Romeley, in Derbyfhire.*

It. is a fulphureous water, impregnated with fea falt.

Its virtues refemble thofe of the *Harro-gate* water.

## S H I P T O N,

*In Yorkfhire.*

It is a fulphureous water, impregnated with fea falt, together with a purging falt.

In its virtues it refembles the *Harrogate* water.

## S O M E R S H A M,

*In Huntingdonfhire.*

This is a chalybeate water, impregnated with green vitriol and allum, and contains alfo fixed air.

The. feafon for drinking it is from M?? to October.

O 4

It is drank in the morning to the quantity
of feveral glaffes.  It is recommended in de-
bilities of the ftomach and bowels, in dyfenteries, hæmoroids, and worms, in nidrous
crudities, in obftruĉtions of the liver
and fpleen, in uterine complaints, in the
ftone and gravel, in the fcurvy, in hyfteric
and hypochondriacal diforders, and many
others.

It is proper to purge before and after the
courfe, and falts may alfo be occafionally
added to it.

Externally it is applied to foul ulcers and
cancers.

### S  P  A  W,

*In the bifhoprick of Leige, Germany, twenty-
one miles South Eaft from the town of
Leige.*

In and about this town there are feveral
fprings, which afford excellent chalybeate
waters : and in Great Britain, they are the
moftdrank of any foreign mineral waters.

The

The principal fprings are,

1. The POHOUN, or POUHON, fituated in the middle of the village.
2. SAUVENIERE, about a mile and a half eaft from it.
3. GROISBECK, near to the Sauveniere.
4. TONNELET, a little to the left of the road to the Sauveniere.
5. WARTROZ, near to the Tonnelet.
6. GERONSTERRE, two miles fouth of the Spaw.
7. SARTS, or NIVESET, in the diftrict of Sarts.
8. CHEVRON, or BRU, in the principality of Stavelot.
9. COUVE,
10. BEVERSE,
11. SIGE,
12. GEROMONT.
} All near Malmdy.

The POUHON, is a flow deep fpring, and is more or lefs ftrong or fpirity according to the ftate of the atmofphere.

It is a chalybeate and acidulous watcr, highly impregnated with fixed air, and co n-tains alfo the foffil alkali, with fome eartlby matter.                                        It

It is in its moſt perfect and natural ſtate in cold, dry weather. It then appears colourleſs, tranſparent, and without ſmell, and has a ſubacid chalybeate taſte, with an agreeable ſmartneſs : at ſuch times, if it be taken out of the well in a glaſs, it does not ſparkle; but after ſtanding awhile, covers the glaſs on the inſide with ſmall air bubbles; but if it be ſhaken, or poured out of one glaſs into another, it then ſparkles, and diſcharges a great number of air bubbles at the ſurface.

In warm, moiſt weather, it loſes its tranſparency, appears turbid or wheyiſh, contains leſs fixed air, and is partly decompoſed. A murmuring noiſe alſo is ſometimes heard in the well.

It is therefore in its greateſt perfection when the weather is cold and dry.

It is colder than the heat of the atmoſphere by many degrees.

It is ſuppoſed to contain the greateſt quantity of fixed air of almoſt any acidulous

lous water; and in confequence thereof has a remarkable fprightlinefs and vinofity, and boils by mere warmth. This, however, foon flies off, if the water be left expofed; though in well corked bottles it is in a great meafure preferved.

It is capable of diffolving more iron than it naturally contains, and thereby becomes a ftronger chalybeate. This is owing to the great quantity of fixed air which it contains,

For the fame reafon an ebullition is raifed in this water on the addition of acids, by reafon that they difengage its fixed air,

It mixes fmoothly with milk, whether it be cold or of a boiling heat.

SAUVENIER. It is of the fame nature with the POUHON water, and feems to be even more acidulous. For it diffolves more iron; ftands longer fine; and preferves its purple tinging quality for a greater length of time,

At

Of the Medicinal Virtues, &c.

At the well it has fomewhat a fmell of fulphur.

GROISBECK. The water is of the fame nature with that of POUHON, but is more acidulous; has a vitriolic tafte, and fomewhat of a fulphureous fmell.

TONNELET. This water feems even to furpafs that of POUHON, and has been too much neglected *. It is one of the moft fprightly waters in the world. It is much colder than either of the Spaw waters; has no fmell; is bright, tranfparent, and colourlefs; and from the rapidity of its motion does not foul its bafon. It has a fmart, fubacid, fprightly tafte, not unlike the brifkeft Champaign wine.

From a variety of experiments it appears, that this water is more ftrongly charged than any of the others with fixed air, on which the *energy* of all waters of this kind depend.

WAR-

---

* The *Tonnelet* fpring, fo much extolled by Dr. Lucas, is, as I am informed by Dr. Simmons, now almoft deferted; and fo is the *Wartroz*.

WARTROZ. Its situation is lowest of any of the springs about Spaw, and is more apt to be foul. But when the well is cleaned out, and the water pure, it is found to be of the same nature as that of POUHON. It is not purgative, as some have afferted.

GERONSTERRE. It refembles the POU-HON water in its brisk, acidulous, and cha-lybeate tafte; and has also a fulphureous fmell at the fountain, which it lofes by being carried to a diftance. This fmell is ftrongeft in warm moift weather.

The air, or vapour, of this water af-fects the heads of fome who drink it, occasioning a giddinefs, or kind of drunken-nefs, which goes off in a quarter or half an hour. The Pyrmont, and feveral other brisk chalybeate waters, are found to have the fame effect.

It bears carriage as well as the other *Spaw* waters, though the contrary opinion has been induftrioufly propagated. It is colder than any of the fprings, the *Ton-nelet* excepted.

It

It does not feem to contain fo great a quantity of fixed air as the *Pohoun*, and fome of the other *Spaw* waters.

SARTS, or NIVESET. It refembles the *Tonnelet* water, but is rather lefs brifk and vi-nous. It is however more acid and ftyptic.

BRU, or CHEVRON. It approaches to the nature of the *Tonnelet* water. The phy-ficians at Leige have artfully decried thi water, becaufe it is not in the principality of Leige. But by every trial it appears not to be inferior to any of the *Spaw* waters.

COUVE and BEVERSE. The *Couve* nearly refembles the *Tonnelet* water; or rather, may be placed in a medium between that and the *Wartroz*. It hardly equals the tranfparency, fmartnefs, and generous vi-nous tafte of the firft, but it greatly fur-paffes the latter. The *Beverfe* agrees with this, only that it does not retain its fmart-nefs fo well by keeping.

LA SIGE. It has fome of the general properties of the *Spaw* waters, but in other refpects it is different.

It

It is moderately fubacid, fmart, and grateful, but has no fenfible chalybeate tafte.

It fparkles like Champaign wine when poured from one glafs to another. Upon ftanding it lofes its fixed air, and throws up a thick mother of pearl coloured pellicle.

It is much more loaded with earthy matters, and lefs impregnated with iron and fixed air than the other Spaw waters.

GEROMONT. As a chalybeate and acidulous water it feems to be nearly of the fame ftrength with the *La Sige*; but it contains a greater quantity of the foffil alkali, together with a mixture of fea falt. The earthy matters, however, are lefs.

*Their Virtues, &c.*—It appears, that thefe waters are compounded of nearly the fame principles, though in different proportions. All of them abound with the mineral elaftic fpirit, or *fixed air*. They contain more or lefs iron, foffil alkali, and calcareous and felenitical earths; together with a fmall a fmall portion of fea falt, and an oily matter common to all waters. Thefe are all kept fufpended, and in a neutral ftate, by means of the aërial acid, or fixed air.

From a review of the contents of thefe waters, it cannot be imagined that their virtues principally depend on the fmall quantity of *folid* matters which they contain. They muft therefore depend moftly on their fubtle mineral fpirit, or *fixed air*. And they are probably rendered more active and penetrating both in the firft paffages, and alfo when they enter the circulation, by means of that fmall portion of iron, earth, falt, &c. with which they are impregnated.

Thefe waters are diuretic, and fometimes purgative; like other chalybeate waters they tinge the ftools black.

They exhilerate and affect the fpirits with a much more kind and benign influence than wine or fpirituous liquors; and their general operation is by ftrengthening the fibres,

fibres. They cool and quench thirft much
better than common water.

They are therefore found excellent in
cafes of univerfal languor or weaknefs, proceeding from a relaxation of the ftomach,
and of the fibres in general, and where the
conftitution has been weakened by difeafes,
or by too fedentary a life. In weak, relaxed, grofs habits; in nervous diforders;
in the end of the gout and rheumatifm,
where the conftitution needs to be repaired;
in fuch afthmatic diforders and chronic
coughs as proceed from too great a relaxation of the pulmonary veffels; in obftructions of the liver and fpleen; in cafes
where the blood is too thin and putrefcent,
occafioned by irregularities, or by fcorbutic
or other putrid diforders; in hyfterical and
hypochondriacal complaints, where the fibres
are too irritable and relaxed, and where the
habit in general needs to be reftored; in
paralytic diforders; in gleets; in the fluor
albus; in fluxes of the belly; and in other
inordinate difcharges proceeding from too
great weaknefs or relaxation of any particular part; in the gravel and ftone; in

P                    female

female obftructions; in barrennefs; and in
moft other cafes where a ftrengthening and
brifk ftimulating refolving chalybeate remedy is wanted; and where there are no
confirmed obftructions, or fo much heat and
fever as to forbid their ufe.

They are, however, generally hurtful in
hot, bilious, and plethoric conftitutions,
when ufed before the body is cooled by
proper evacuations. They are alfo hurtful
in cafes of fever and heat; in hectic fevers
and ulcerations of the lungs, and of other
internal parts, particularly where there is
no free outlet to the matter; and alfo in
moft confirmed obftructions attended with
fever.

The ufual feafon for drinking them is in
July and Auguft, or during the fummer
months from May to September. The
water, however, is beft in the winter, and
in dry, frofty weather; and probably might
then be drank to greateft advantage.

If they lie cold on the ftomach, a few
carraway feeds, cardamoms, or other aro-
matic,

matic, may be taken with them. The veſſel out of which it is drank may alſo be warmed with hot water, or a little warm water may be added immediately before drinking. It muſt always be drank before noon.

The quantity to be drank ſhould be different according to the age, conſtitution, and other circumſtances of the patient. The only certain rule is, that quantity which the ſtomach can bear without heavineſs or uneaſineſs. The greater the quantity any one drinks, the better, provided it agrees, and paſſes well off. It is adviſeable to begin with drinking a glaſs or two ſeveral times in a day, encreaſing the quantity daily, as far as the ſtomach will bear. To continue that doſe during the courſe, and to finiſh by leſſening it by the ſame degrees by which it was augmented. Moderate exercife is proper after drinking. It is to be continued for ſeveral weeks or months, according to the circumſtances.

Previous to the uſe of the water, it is proper to cleanſe the firſt paſſages by gentle purges, and if judged neceſſary, an emetic

alſo

alfo fhould be given. During the courfe, likewife coftivenefs fhould be prevented, by occafionally adding Rochel falts, or rhubarb, to the firft glaffes of water in the morning.

When there is too much heat, the faline draughts, nitre, vegetable acids, or the like, fhould be given, and the elixir of vitriol has been added to the water, in intermittent feverifh complaints, with good effect.

A cooling regimen fhould be obferved while drinking thefe waters, as alfo regular hours, and quietnefs, or chearfulnefs of mind.

In cafes of rigidnefs of the fibres, the warm bath is recommended, among the beft preparatives to a courfe of thefe waters ; and, hence bathing at *Aix-la-Chapelle,* or at *Chaude Fontaine,* is often premifed. The warm bathing may occafionally be repeated during the courfe. In oppofite cafes, the cold bath is recommended.

The Spaw water is ufed alfo externally,

. in

in a variety of cafes, with good fuccefs. It is ufed as an injection in the fluor albus, and in ulcers and cancers of the womb, and alfo in the gonorrhœa; it is ferviceable in venereal aphthæ, and ulcers in the mouth. It is ufed to wafh phagadenic ulcers : it is recommended by way of gargle for relaxed tonfils, and for faftening loofe teeth; it is alfo good in other relaxations; and it is faid to cure the itch, and fimilar complaints, by wafhing and bathing; an internal courfe being alfo obferved at the time.

As the Spaw waters are impregnated with different proportions of the fame ingredients, they may be chofen differently, according to the intentions we have in view. The *Pohoun* is the ftrongeft chalybeate. The *Tonnelet* and *Geronflerre*, are weaker chaly- beate, but are brifker, and rather more fpi- rituous. The *Groefbeck*, *Sauveniere*, and *Wartroz*, are ftill weaker chalybeates, but highly impregnated with calcareous and fe- lenitical earths, and contain alfo a greater proportion of the foffil alkali. The *Gero- mont*, is likewife a weak chalybeate, but contains a great deal of calcareous and fele-

P 3 nitical

nitical earth, and about three times as much alkaline falt as any of the others. The four laft waters, therefore, will be better in diforders arifing from an acid caufe, and as diuretics, particularly the *Geromont*.

## S T E N F I E L D,

### *In Lincolnfhire.*

It is a chalybeate laxative water, and refembles that of *Orfton*. It is light, clear, pleafant tafted, and full of fpirit at firft, but on long ftanding in its large refervoir fpoils.

## S T R E A T H A M,

### *In Surry, near London.*

The water has a yellowifh tinge, and throws up a fcum variegated with blue, green, and yellow. Its tafte is fomewhat faline and difagreeable.

It is a mild purging water, and may be drank to the quantity of three or four pints.

It is alfo diuretic, and is faid to be found ufeful in diforders of the eyes.

STANGER,

# S T A N G E R,

*Near Cockermouth, in Cumberland.*

This is a falt chalybeate, or vitriolic water; and, when drank to four or five pints, operates with violence both upwards and downwards.

# S T O K E.

See *Jeſſop's Well.*

# S U C H A L O Z A,

*About a mile from Hungarian Broda, in Germany.*

It is an acidulous water, reſembling that of *Nezdenice* in virtues.

It is greatly eſteemed in the neighbourhood for the cure of ſcrophulous and other diſorders, in which waters of this kind are ſerviceable; and is drank with victuals inſtead of ſmall beer and wine.

# S U T T O N B O G,

*In the county of Oxford, near to Northamptonſhire.*

This is one of the waters termed *ſulphureous.*

P 4 It

It has an intolerable fœtid fmell, like rotten eggs. Its tafte is faltifh and pungent, like foap lees.

It throws up a blue fcum, and the mud at the bottom is jet black. In half an hour it turns filver of a copper colour.

It contains an alkaline falt, together with a little fea falt.

It is a mild laxative, or purging water.

It is ufed both for drinking and bathing; and ulcers, tumours, fcrophulous, and other difeafes of the fkin are fuccefsfully wafhed with it. The mud is alfo made ufe of.

## SWADLINGBAR,

*In the county of Cavan, Ireland.*

The water is fometimes tranfparent and colourlefs; at other times fomewhat whitifh.

It has a ftrong fulphureous fmell, which it retains long in bottles well corked. It tinges filver of a blackifh or copperifh colour.

The

The well is commonly covered with a whitifh or bluifh fcum ; and depofits a mud which burns, on the red hot iron, with a blue flame.

It contains the foffil alkali, together with a little fea falt and earth.

It refembles in its virtues the water of *Drumgoon.*

## SWANSEY,

*In Glamorganfhire, North Wales.*

It is impregnated with green copperas, and therefore is of the nature of the *Shadwell* water.

Taken inwardly it is alfo faid to ftop purgings ; applied outwardly it ftops bleeding.

## SYDENHAM,

*In Kent, near London.*

The water is fomewhat bitterifh to the tafte.

It is purgative, and of the nature of *Epfom* water, but weaker.

TAR-

# TARLETON,

*Eight miles from Preſton in Lancaſhire.*

This is a chalybeate water, and drank to the quantity of three or four pints proves purgative. In its virtues it ſeems to reſemble the *Scarborough* water. It has a ſomewhat ſulphureous ſmell when firſt drawn.

# TEWKSBURY,

*In Glouceſterſhire.*

It is a purging water, of the nature of thoſe of *Aĉton, Pancras,* and *Epſom.*

There are two other ſprings of the ſame kind in the neighbourhood; one of them is in Walton grounds, the other in Teddington grounds,

# THETFORD,

*In the county of Norfolk.*

This is a chalybeate and acidulous water, and contains alſo the foſſil alkali.

It operates by urine, and alſo gently by ſtool.

It

It is recommended in pains of the ftomach and bowels; in lofs of appetite; in relaxed ftate of the fibres; in hyfteric diforders; and in beginning confumptions.

## THOROTON,

*Near Newark upon Trent, Nottinghamfhire.*

It is a chalybeate laxative water, refembling that of *Orfton.*

## T H U R S K,

*In the North Riding of Yorkfhire.*

It is a brifk, fparkling, chalybeate water, and is alfo purgative and diuretic. It refembles the *Scarborough* and *Cheltenham* waters.

## TIBSHELF,

*In Derbyfhire.*

This is a fine clear chalybeate; and when poured from one glafs to another, fparkles like the *Spaw* water, which it refembles in virtues.

TIL-

# T I L B U R Y.

*The spring which affords this water is situated near a farm house at West-Tilbury, near Tilbury-Fort, in Essex.*

This water is not quite limpid at the well, but is somewhat straw-coloured.

It is soft and smooth to the taste; though after being agitated in the mouth, it leaves a small degree of roughness on the tongue.

It throws up a scum variegated with several colours, which feels greasy; and effervesces with spirit of vitriol.

It mixes smooth with milk, but curdles with soap. When boiled it turns milky; a fourth part of mountain wine fines it immediately; and all acids do the same.

It operates chiefly by urine; though it is also somewhat purgative; and increases perspiration.

It is in esteem for removing glandular obstructions. It is good in bloody fluxes, purgings, and the like. In diforders of the ftomach arifing from acidity; in the gravel; fluor albus; and immoderate flux of the menfes.

As a diuretic it is good in dropfical complaints.

It gently warms the ftomach; ftrengthens the appetite; and promotes digeftion. It is alfo of fervice in lownefs of fpirits. From its efficacy in removing obftructions of the glands, it is recommended in fcurvies and cutaneous difeafes; and its virtues in thefe complaints feem to be confirmed by the tingling which it occafions in the fkin.

The dofe is ufually a quart in a day.

The water is fuppofed to owe its virtues' to a native alkaline falt, which may be obtained from it by evaporation, and to its *fixed air.*

## T O B E R  B O N Y,

### *In Ireland.*

This fpring is fituated about four miles north of Dublin.

The

The water is fweet, and foon lathers with foap.

Before rain and wind it yields a fetid fmell. Its fediment, when placed on hot iron, turns black and fetid.

It contains an alkaline falt, together with a calcareous earth, and an oily or bituminous matter.

Its virtues are fimilar to thofe of the *Tilbury* water, but in a lefs degree.

## T O N S T E I N.

*In the Biſhoprick of Cologne, Germany.*

This is among the moſt noted waters of Germany.

The water has a briſk fubacid taſte, at the fountain, which is loſt by expofure to the air.

It is clear and limpid when taken up from the well, but becomes turbid by ſtanding; owing to the lofs of its fixed air.

It contains an alkaline falt, together with a little chalky earth, and fea falt.

Its

Its virtues are fimilar to thofe of the *Selt-zer* waters, but it is more purgative.

It may alfo be ufed with advantage for common drink, either by itfelf or mixed with wine; and that either in acute or chronic difeafes, where diuretic or deobftruent reme-, dies are required.

# T O W N L E Y,

See *Hanbridge.*

# T R A L E E,

*In the county of Kerry, Ireland.*

It is a chalybeate water, of the nature of. that of *Caftleconnel.*

# T U N B R I D G E.

*The* WELLS *are fituated about five miles from the town of Tunbridge, in Kent.*

This is at prefent one of the moft famous chalybeate waters in England, and the moft reforted to of any, though it does not feem to be preferable to many others in this king-dom.

It

It is a brifk, light water, has a ferruginous tafte, and contains alfo a little fea falt.

Expofed to the air it foon lofes its virtues; as it does alfo in a few days in bottles.

It is ufual at times to mix with the firft glafs of the water, taken in the morning, either a little common falt, or fome other purging falt, in order to make it operate by ftool. If the ftomach be foul, it is apt to vomit.

It is chiefly reforted to in June, July, and Auguft; and is recommended in all thofe diforders in which the celebrated *Spaw waters of Germany* are ferviceable. It poffeffes the fame general virtues as thofe waters, but in a lefs degree.

## U P M I N S T E R,

*Near Brentwood, in Effex.*

This is a ftrong fulphureous water, impregnated with a purging falt, and the foffil alkali.

It is purgative and diuretic; and in its virtues feems to refemble the *Afkeron* water.

V A H L S,

## V A H L S,
*In France.*

The well is near Vahls, in Dauphiny.

The water has a brifk fubacid tafte at the fpring ; which is loft before it reaches Paris, for it then taftes faltifh.

It contains the foffil alkali.

It is diuretic, and fomewhat purgative ; and is fimilar in virtues to the *Seltzer* and *Clifton* waters, though lefs powerful.

N. B. Near to this is another fpring, called *La Marie,* but weaker.

## W A R D R E W,
*In Northumberland.*

It is fituated between Cumberland and Northumberland, on the banks of the river Arden.

It is the moft cold fulphureous water in the three northern counties. It contains alfo fea falt, and therefore refembles in vir-tues the *Harrogate* water.

Q                    It

It lofes both its fmell and virtues by carriage and keeping.

## WEATHERSTACK,

*In Weftmoreland.*

This is a weak chalybeate water, but contains a large portion of fea falt. In the
fummer it fmells of fulphur, but not in the
winter.

It is purgative; and the dofe is two or
three pints.

## WELLENBROW,

*In Northamptonfhire.*

It is a light chalybeate water, refembling
that of *Iflington*.

## WEST ASHTON,

*In the parifh of Steeple Afhton, Wiltfhire.*

It is a weak chalybeate water, refembling
thofe of *Iflington* and *Tunbridge*.

## WESTWOOD,

*Near Tanderfley, in Derbyfhire*

This is a vitriolic chalybeate, fomewhat
refembling the *Shadwell* water.

It

It is recommended externally for old fores in the legs.

N. B. The coal waters, in general, in this part of the country, are alfo vitriolic.

## W E X F O R D,
### *In Ireland.*

It is an agreeable chalybeate water, fimilar in virtue to that of *Iflington.*

## W H I T E - A C R E,
### *Near Trales, in Lancafhire.*

This is a very clear, brifk chalybeate water, refembling that of *Lancafter* in virtues, but it is faid rather to bind than purge.

## W I G A N,
### *In Lancafhire.*

It is a clear chalybeate water, refembling thofe of *Hampftead* and *Iflington.*

## W I G G L E S W O R T H,
### *In the parifh of Long Prefton, in the Weft Riding of Yorkfhire, four miles South of Settle.*

The water is very black, and has a ftrong fulphureous fmell, with a faltifh tafte.

Q 2                    Drank

Drank to the quantity of three quarts, it purges, and two quarts are faid to vomit, though it is rather uncommon, that more fhould be required for the latter than the former.

## W I L D U N G A N,

*In the country of Waldeck, Germany.*

This water at the fountain, has a brifk fubacid tafte, which it lofes by expofure.

It is of the fame kind with that of *Buch*, but weaker.

It is one of the mildeft acidulæ known, and may be ufed as common drink alone, or mixed with a fmall portion of wine.

Though it is not efteemed ftrong enough to remove obftinate chronic difeafes, and clear the firft paffages, yet it is excellent for blunting and dilating acrid, fcorbutic, and gouty humours, when taken in large quantity, and for a fufficient length of time.

W I T H A M,

# W I T H A M,

*In Effex,*

This is a chalybeate water of confiderable ftrength, and is alfo impregnated with fea falt, but it will not bear carriage, and muft be drank at the fountain.

It is very diuretic, and has been fuccefs-fully prefcribed in hectic fevers, in weaknefs occafioned by long difeafe, in lownefs of fpirits, nervous complaints, want of appetite, indigeftion, habitual cholic, and vomiting; in agues, in the jaundice, and beginning dropfy; in the gravel, and in afthmatic and fcorbutic diforders.

# W I R K S W O R T H,

*In Derbyfhire.*

It is a weak fulphureous water, impregnated with a purging falt, and is alfo chalybeate.

It is recommended in fcrophulous, and cutaneous diforders.

ZA-

## Z A H O R O V I C E,

*In Germany.*

The fpring is near to this village, in the diftrict of the Caftle of Suietlovia, in a rocky valley, by the fide of the river Nezdenice.

It is an acidulous water, falter, but lefs acid than that of *Nezdenice;* and it is alfo fomewhat pungent and fœtid.

It is in great efteem in the neighbour-hood, particularly for the cure of fcrophu-lous diforders.

C O N-

# CONCLUSION.

FOR the fake of brevity, I have omitted a particular defcription of each water in the preceding account, and occafionally referred the reader to fome water of the fame kind which has been more fully treated of; and the general virtues of the different claffes of waters are alfo defcribed at large in the Introduction.

In the Appendix to Dr. Prieftley's tract, I have given directions for imitating fome of thofe waters. The acidulous waters of the 5th clafs, for example, may be imitated, and even excelled, by fimply impregnating water with *Fixed Air*. The folid ingredients are known to be of little or no confequence. If, however, thefe are defired, they may be added in the proportions directed under the article *Seltzer water*; though it is by no means neceffary that thofe proportions fhould be ftrictly adhered to.

A Purging water, anfwering perhaps all the intentions of thofe of the 6th clafs, may be made as directed for the *Seidfcutz water*. Rochelle

Rochelle or Glauber's falt may be fubftituted for the Epfom, if the latter be too naufeous ; and a little common falt may alfo be added. If the water to be imitated be a falt water, like that of the fea, the common falt fhould be in the greater proportion.

The chalybeate waters of the 1ft clafs may be elegantly fubftituted, by water impregnated with Fixed Air, in which ironfilings, or wire, has been infufed : or they may be made as directed under the articles *Spaw* and *Pyrmont water*. The chalybeate purging waters of the 2d clafs may be imitated by adding to a gallon of this water two or three ounces of Epfom, or other purging falt, and, if you will, a little fea falt.

For the fulphureous waters of the 3d clafs water impregnated with fulphureous air may be advantageoufly ufed: or they may be made as directed under the article *Aix-la-Chapelle water*. If they are alfo required to be chalybeate, or purging, or both, iron-filings, or Epfom falt, or both thefe may be added, together with a little fea falt, according to circumftances. For cold fulphureous waters both fixed and fulphureous airs are to be employed; as mentioned in the Appendix ;

and

and even for the hot fulphureous waters it may be proper to put a fmall proportion of chalk with the liver of fulphur into the lower veffel A of the apparatus.

They who have a knowledge of natural philofophy, will perceive that thefe artificial waters are not only equal, but even fuperior to the natural ones, efpecially when they cannot be drank at the fpring head. Their virtues, for the moft part, depend on their *volatile* principles, and art can make water imbibe more than double the quantity of fixed, or fulphureous air, that the ftrongeft natural waters are ever found to contain. The latter are alfo frequently impregnated with hurtful or, at leaft, ufelefs ingredients; and we cannot always be fure that we have them genuine. It is not, however, by any means, the Author's wifh to profcribe the ufe of the natural waters. Many of them have particular virtues, as has been proved by undoubted experiments: and there are others which art cannot yet fufficiently imitate.

Many people again, through prejudice, will not ufe the artificial waters, as they do not believe it poffible that they can be made fufficiently to refemble the natural ones; but

even.

even thofe will not object to the ufe of water
impregnated with *fixed* or *fulphureous air*
in a *medicinal* view.

Water impregnated with fixed air is now
known to be a very powerful antifeptic, or
corrector of putrifaction. It will preferve
flefh kept in it fweet, and even reftore it
after it becomes putrid. It is therefore given
with great fuccefs in putrid fevers, in the fea
fcurvy, in dyfenteries, in mortifications, and
in other diforders arifing from a putrid caufe,
or attended with putrifaction, a draught of
it being taken now-and-then, or even by
way of common drink. But the ingenious
Mr. Bewly has invented a ftill better method
of exhibiting fixed air, as a medicine : he
directs a fcruple of alkaline falt to be diffolved in a fufficient quantity (fuppofe a
quarter of a pint, or lefs) of water, which
is to be impregnated with as much fixed air
as it can imbibe ; this is to be drank for
one dofe. ‡ If immediately after it a fpoonful
of lemon juice, mixed with two or three
fpoonfuls of water, and fweetened with fugar, be drank, the fixed air will be extricated in the ftomach ; and by this means a
much

‡ Mr. Bewly directs it to be prepared in larger quantity at a time, (as indeed it ought, in order to fave
trouble) and calls it his *Mephitic Julep*.

much greater quantity of it may be given than the fame quantity of water alone can be made to imbibe. In this way I have given it in the above diforders, as well as in thofe that follow, with the beft effect.

Fixed air acts as a corroborant; and therefore may be given with fuccefs in weaknefs of the ftomach, and in vomitings arifing from that caufe.

It has already been noticed, in the Introduction, that if mild calcareous earth be fuperfaturated with fixed air, it becomes foluble in water. The calculus, or ftone in the bladder, confifts partly of this earth. Fixed air therefore has been given as a *folvent* in this cafe with fuccefs.

When the lungs are purulent, fixed air mixed with the air drawn into the lungs, has repeatedly been found to perform a cure.

The bark may be given with advantage in water impregnated with fixed air, as they both coincide in the fame intention.

Fixed air may be applied by means of a fyringe, or otherwife, to putrid ulcers, mortified parts, ulcerated fore throats, and in fimilar cafes, and it has been found to have remarkable efficacy. It may alfo be given internally at the fame time.

In putrid dyfenteries, and in putrid ftools, fixed air may be given by way of clyfter, as hath been obferved by the learned and ingenious Dr. Prieftley (whom I have the honour to call my friend) in the former part of this work. Fermenting cataplafms are of fervice chiefly as they fupply fixed air to the part.

In cafes of putridity, fixed air has been fuccefsfully applied to the furface of the body, expofed to ftreams of it. And there are other cafes in which it has been found ferviceable. It is alfo an excellent cooling as well as ftrengthening beverage in hot relaxing weather, and it has befides the advantage of being pleafant tafted.

The virtues of water impregnated with *fulphureous air* may be collected from what was faid in the Introduction, concerning fulphureous waters.

F I N I S.

www.ingramcontent.com/pod-product-compliance
Lightning Source LLC
Chambersburg PA
CBHW021527210326

41599CB00012B/1411